Field Manual
No. 100-8

Headquarters
Department of the Army
Washington, DC, 24 November 1997

The Army In Multinational Operations

Contents

Preface

This manual is a guide for Army commanders and staffs operating in a multinational environment across the full range of military operations. This doctrine for multinational operations takes into account each type of operation and varying political objectives, force compositions, operating areas, and other factors. It also considers general similarities in most multinational operations. Because of the varying compositions and varying political objectives, consensus is extremely difficult to obtain for multinational operations. Harmonization of the multinational force is critical to ensure unity of effort.

FM 100-8 blends key points of Joint Pub 3-16 into its approach to ensure consideration by Army elements of a joint force. This manual addresses the Army's roles and functions within a multinational operation; multinational leadership; command relationships, to include examples of possible command relationships; and, finally, the key functional planning considerations of the multinational commander at the operational and tactical levels of war. The combat functions listed in FM 100-5 provide a guideline for this manual's content. FM 100-8 should be read in conjunction with—

- CJCSM 3500.04.
- Chapter 5 of FM 100-5.
- FM 100-23.

The primary audiences for FM 100-8 are Army forces (ARFOR), Army service component commanders (ASCCs), and other senior leaders and their staffs. The focus is on ARFOR and ASCC responsibilities for conducting operations as part of a multinational force. Information contained herein will help other national forces and other services plan and conduct multinational operations with US Army forces. This doctrine will also assist Army branch schools in teaching multinational army operations.

The proponent for this publication is HQ TRADOC. Send comments and recommended changes directly to Commander, US Army Combined Arms Center, ATTN: ATZL-SWW, Fort Leavenworth, KS 66027.

Unless this publication states otherwise, masculine nouns or pronouns do not refer exclusively to men.

Introduction

Where commonality of interest exists, nations will enter into political, economic, and/or military partnerships. These partnerships will occur in both regional and worldwide patterns as nations seek opportunities to promote their mutual national interests or seek mutual security against real or perceived threats. Cultural, psychological, economic, technological, and political factors all influence the formation and conduct of alliances and coalitions.

America's interests are global, but its focus is regional, and existing alliances and past coalitions reflect that focus. Alliance participants establish formal, standard agreements for broad and past coalition objectives. Alliance nations strive to field compatible military systems, structure common procedures, and develop contingency plans to meet potential threats. As forces of these nations plan and train together, they develop mutual trust and respect. Present alliances and agreements include the North Atlantic Treaty Organization (NATO); the American, British, Canadian, and Australian (ABCA) Armies Standardization Program; defense and cooperation treaties with Korea and Japan; and the Inter-American Treaty of Reciprocal Assistance.

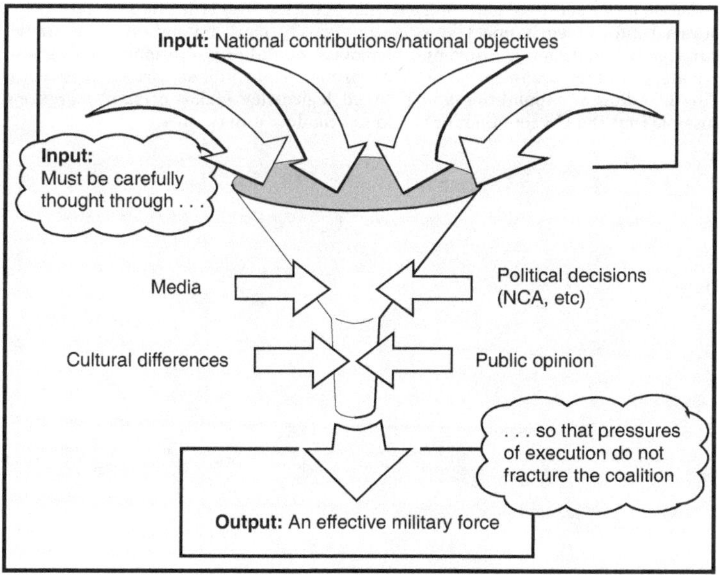

Figure I-1. Coalition Building

While alliances form the basis for responding to a variety of regional threats, new coalitions, such as Operation Desert Shield in August 1990, emerge to meet future national strategic requirements. Coalitions, which are created for limited purposes and for a limited length of time, do not afford military planners the same political resolve and commonality of aim as alliances. Thus, planners must closely study the political goals of each participant as a precursor to detailed planning. Political considerations weigh more heavily with coalitions than with alliance operations.

Multinational military operations are not new. Most major military operations in the twentieth century have been both joint and multinational: World War I, the Allied Intervention in Russia, World War II, the Korean Conflict, the Vietnam Conflict, and various UN-sponsored operations. Since human nature has not changed, regional conflicts over territory, religion, politics, and economics, such as those that prompted previous military operations, will continue to be widespread. The precise role of land forces in these operations will vary according to each political and military situation. US Army participation is likely for three reasons: only land forces can hold terrain and control populations, Army structure contains unique capabilities other services do not have, and soldiers on the ground are a clear demonstration of political resolve.

The United States' strategic principal of collective security caused the United States to join several alliances and form coalitions. This requires the Army to conduct multinational military operations with forces from other nations. These operations will generally include a variety of governmental and nongovernmental agencies, other services, and international agencies. The other reason the United States conducts such operations is that rarely can one nation go it alone. Participating national contingents, nongovernmental organizations (NGOs), and private voluntary organizations (PVOs) bring certain unique core competencies. This blending of capabilities and political legitimacy makes certain operations possible that the US could not or would not conduct unilaterally.

Chapter 1

Fundamentals

Politics can have a fundamental impact on military operations, particularly when the cooperation of many nations is required for success. The interaction of multiple Clausewitzian trinities of governments, peoples, and militaries creates a changeable and fragile partnership that can be fractured if the interests of individual nations are threatened.

Wayne Danzik, *Coalition Forces in the Korean War*

Even the soldiers of a Democracy cannot always understand the reasons back of strategic situations. Political and military reasons are worked out in cabinets and general staffs and soldiers obey orders.

Newton D. Baker, Secretary of State under President Woodrow Wilson

Multinational operations include military forces from different nations. These operations are normally interagency operations. Conducting operations with foreign military partners, like operations with civilian partners, is uncommon to many soldiers, so a clear understanding of this different environment is necessary. This chapter provides the fundamental background that units will face when conducting multinational operations. US forces are increasingly more involved in United Nations (UN) operations and must plan accordingly.

The majority of US peace operations will be part of a UN peace operation. Their multinational character merits particular attention because national interests and organizational influence may compete with doctrine and efficiency. Consensus is painstakingly difficult, and solutions are often national in character. Commanders can expect contributing nations to adhere to national policies and priorities, which at times complicates the multinational effort.

UN-sponsored operations normally employ a force under a single commander. The secretary general appoints the force commander with the consent of the UN Security Council. The force commander reports either to a special representative of the secretary general or directly to the secretary general. While the force commander conducts day-to-day operations with fairly wide discretionary powers, he refers all policy matters to the special representative or secretary general for resolution. FM 100-23 is an excellent reference for UN operations.

OPERATIONS

Multinational operations is the overarching collective term to describe military actions conducted by forces of two or more nations, typically within the structure of a coalition or alliance. If that operation takes place within the NATO or Combined Forces Command (CFC) in Korea, it is an *alliance operation*. If it is a multinational action outside the bounds of an established alliance, it is a *coalition operation*. These terms should not be used interchangeably. See the glossary for exact definitions.

DOCTRINE

Multinational operations can have uniform, robust, and widely accepted doctrine, such as Allied Joint Publication (AJP)-1 for NATO; however, this may not always be the case. Each situation and combination of forces is unique to a varying extent. National forces operate in their normal manner internally, and their actions generally conform to the overall direction of the multinational force (MNF).

Each alliance develops its own common doctrine, tactics, techniques, and procedures (TTP), but these normally conform with each member nation's doctrine. Coalitions are much less structured, and commanders must know the differences in the other partners' goals, doctrine, organizations, equipment, terminology, cultures, religions, and languages. For all multinational operations, commanders and staffs must know common considerations and factors beforehand to conduct the operations efficiently.

US Army forces operating as part of a multinational military command (alliance or coalition) should follow multinational doctrine and procedures previously agreed to by the US. For doctrine and procedures not ratified by the US, commanders should follow the multinational command's doctrine and procedures, where applicable. US joint doctrine provides the US position for multinational doctrine development consistent with existing security procedures.

PRINCIPLES

The principles of war apply to multinational operations. Two principles are key to successful multinational operations, *objective* and *unity of command.*

OBJECTIVE

Commanders must focus significant energy on ensuring that all multinational operations are directed toward clearly defined and commonly understood objectives that contribute to the attainment of the desired end state. No two nations share exactly the same reasons for entering into a coalition or alliance. Furthermore, each nation's motivation tends to change during the situation. National goals can be harmonized with an agreed-upon strategy, but often the words used in expressing goals and objectives intentionally gloss over differences. Even in the best of circumstances, nations act according to their own national interests. Differing goals, often unspoken, cause each nation to measure progress differently. Thus, participating nations must agree to clearly defined and mutually attainable objectives.

UNITY OF COMMAND

The principle of unity of command is better explained as unity of effort for multinational operations. Successful multinational operations center on achieving unity of effort. This unity is extremely difficult to achieve without each commander's individual attention. There must be a certain spirit of cooperation among the nations, and each participating nation must agree to provide the MNF commander sufficient authority to achieve unity of effort. This authority, however, is seldom absolute, and consensus is critical to success.

BATTLE DYNAMICS

Battle dynamics apply to multinational operations, and the details of their application are covered throughout this manual. This paragraph discusses battle dynamics in a multinational context.

Battle Command

Leaders must assimilate a massive volume of information to visualize the battlespace, assess the situation, and direct the military action required to achieve victory. Visualizing the battlespace or disaster area requires the commander to go where his presence can be felt, where he can best influence operations, and where he can best express his will. Personal and interpersonal communication skills among differing nationalities are even more important in multinational operations. Even within US forces, electronic passage of communications will not serve as a substitute for face-to-face meetings. In multinational operations, language differences, national pride, and mutual confidence make interpersonal communication skills even more critical. Leaders must understand and accommodate national customs and more. Decision making and leadership are more complex in multinational operations for the same reasons. The commander must review command and control (C^2) techniques to ensure he gets accurate input and his desires are properly communicated and executed.

Battlespace

Within a given battlespace, commanders must understand the effects of geography, infrastructure, and terrain; appropriately apply organic capabilities; and integrate joint, interagency, and multinational assets. Unity of effort is essential to operations within a given battlespace. Assets should be employed to maximize effects with less regard to national ownership. Defining a force's battlespace requires a detailed understanding of the other forces' weaponry, sensors, and methods of employment. Synchronization is more difficult because of the extra friction caused by different languages, doctrine, and procedures.

Depth and Simultaneous Attack

The US can expect some disparity between the capabilities of its forces and those of its partners. Where the US provides a preponderance of forces, it will probably control or at least coordinate most attacks, both of lethal and nonlethal means. Extra effort may be necessary to synchronize the ground maneuver of multinational partners with US deep strikes. Clear control measures are needed to avoid fratricide.

Early Entry, Lethality, and Survivability

Although most recent early entry operations conducted in a multinational setting have been unopposed, this may not always be the case. The key points are coordinated with host nation (HN) forces and other multinational partners. Timely coordination enables planners to tailor forces and make the best early entry decisions regarding the flow of forces. Other nations can provide many core capabilities that the US may not need to deploy; conversely, if other forces do not have key capabilities, US planners can move such capabilities up in the deployment flow.

Combat Service Support

Multinational logistics are crucial, especially during deployment, entry, and redeployment operations. Support functions must be integrated among different national military forces, HN assets, and contractors. Chapters 3 and 4 discuss these issues in detail.

MUTUAL CONFIDENCE

US commanders throughout our history have operated with MNFs, starting with George Washington and his partnership with the French allies. In the twentieth century, Generals Pershing, Eichelberger, Eisenhower, Ridgway, Westmoreland, Abrams, and Schwartzkopf and Major General Nash all operated in difficult multinational environments. Success came from the commander's focus on the political objective, assigned mission, patience, sensitivity to the needs of other coalition members, and a willingness to compromise or come to a consensus when necessary.

Tangible Considerations

After World War II, General Eisenhower said that "mutual confidence" is the "one basic thing that will make allied commands work." This is true for all multinational operations. This mutual confidence stems from a combination of tangible actions and entities and intangible human factors. Chapters 2, 3, 4, and 5 cover the tangible considerations, such as liaison and language.

Intangible Considerations

The intangible considerations that must guide the actions of all participants, but especially the senior commander, are briefly described here. These factors cannot guarantee success, but ignoring them can almost always guarantee failure.

Rapport

Rapport must be established among all of the nations' senior commanders. Rapport is a personal, direct relationship that only the commanders can develop. The commander's first concern when establishing rapport is the complete analysis and understanding of the characteristics, personality, capabilities, ambitions, and personal and professional habits of those various senior commanders within the MNF. Once he has this understanding, the keys to developing rapport are respect, trust, patience, and the ability to compromise. The result of good rapport among leaders is successful teamwork by their staffs and subordinate commanders and overall unity of effort.

Multinational operations are often difficult. Differing national agendas can be disruptive, but, on a more personal level, the natural competitiveness among soldiers and nations can become a serious problem. Such competitiveness can be a motivating factor if properly managed, but, left unchecked, it can become destructive to the cohesion of the multinational effort. Multinational commanders at all levels must reinforce the fact that all MNFs are on the same team. Establishing an atmosphere of cooperation and trust at the highest levels of any multinational effort is essential. When such an atmosphere is established, subordinate commands are influenced positively.

HISTORICAL EXAMPLE

During the New Guinea campaign in World War II, General Douglas MacArthur dispatched Lieutenant General Robert Eichelberger to the area to improve the poor situation. Prior to Eichelberger's arrival in Papua, an adversarial relationship had developed between US and Australian commanders. Australian leaders had questioned the fighting capability of US forces and similar charges were made by the American side regarding the Australians.

One of Eichelberger's first acts upon assuming command in Papua was to send a communique to Lieutenant General Edmund Herring, the Australian appointed by MacArthur to be the field commander of the Papua operation. In this message, Eichelberger informed Herring that he was anxious to cooperate with the Australians. Herring, who had been terse in his previous correspondence with Elchelberger, mellowed considerably subsequent to this event.

Upon Herring's return to Port Moresby to command all operations on New Guinea, he appointed Eichelberger field commander of a corps comprised of both Australian and American units. This was the first time in the war that the Australians placed their troops under the tactical command of an American officer. The quality Eichelberger possessed that made him particularly suited to the multinational environment was his ability to demonstrate to his allies his belief that personal and national prestige cannot be allowed to interfere with harmony within the partnership. These abilities, coupled with his skill in promoting a spirit of teamwork with allied forces, enabled him to command effectively in the multinational arena.

The commander must be visible to members of the coalition. Personal visits to allied units provide the opportunity to assess capabilities, readiness, and morale. The commander must ensure equitable treatment and exposure of all units, regardless of national background. Failure to do so may be perceived as prejudice and result in political repercussions. All members must have fair representation on multinational planning staffs to preclude allegations that any nation was excluded from participation in the decision-making process.

Establishing rapport within the partnership is more easily accomplished when nations combining forces share similar cultural backgrounds. Conversely, the partnership can be fractured when members come from diverse cultural backgrounds and do not attempt to respect one another's cultural sensitivities.

Desert Storm proved that a successful coalition among nations having widely disparate social and cultural norms is achievable. American commanders ensured that their soldiers received education regarding social norms prevalent in the Middle East and imposed measures to respect these norms. Arabic coalition leaders, for their part, allowed American and other Western coalition forces liberties that would not normally be permitted under Islamic law. Allowing female soldiers to drive vehicles and perform other duties forbidden by Saudi Arabian law and permitting the conduct of non-Islamic religious services would not be considered major concessions by Western standards, but in Saudi Arabia, these allowances were extremely significant.

Respect

Respect must exist among multinational partners. All should perceive they are contributing equally toward accomplishing the mission, regardless of the senior member's rank or size of the national force. Respect for the partners' culture, religions, customs, and values, combined with understanding and consideration of their ideas, will solidify the partnership. Without the genuine respect of others, rapport and mutual confidence will not exist, and lack of respect may lead to friction, which jeopardizes mission accomplishment.

Knowledge of Partners

Know the multinational partners as well as you know the enemy or adversary. Much time and effort is expended in learning about the enemy and perhaps even more important is a similar effort to understand the partners. Each partner in an operation has a unique cultural identity. Nations with similar cultures face fewer obstacles to interoperability than nations with divergent cultural outlooks. Commanders and planners must learn the capabilities of partner nations or organizations. These capabilities differ based upon national and organizational interests and objectives, political guidance and limitations on the national force, doctrine, organization, rules of engagement (ROE), rules of interaction, laws of armed conflict (LOAC), equipment, religions, customs, history, and a myriad of other factors.

HISTORICAL EXAMPLE

In Somalia, certain Somali factions targeted the Nigerian national contingent for repeated attack. The Nigerian sector did not have any unique features warranting attack, and the Nigerians did not provoke attacks. What planners had not considered was that Nigeria had earlier granted the former President of Somalia, Siyad Barre, political asylum in Nigeria. Some Somalis took revenge on Barre by targeting the Nigerian soldiers for attack.

Appropriateness of Missions

All participants must perceive missions as appropriate, achievable, and equitable in terms of burden and risk sharing. Capabilities are an obvious factor, but national honor and prestige may be as important to the partnership as battlefield capability. Partners should be included in the planning process, and their opinions must be sought concerning mission assignment. The political impact of high casualties must be balanced against practical military choices.

Patience

Developing effective partnerships takes time and attention. Diligent pursuit of a trusting, mutually beneficial relationship with multinational partners requires untiring, even-handed patience.

COMMUNICATIONS

The ability to communicate in the different partners' languages cannot be overemphasized because language barriers represent a significant challenge. Language is more than the direct translation of words. Much information is conveyed by word choice, mannerisms, and so forth. Language capability speeds battle command, reduces confusion, and contributes to mutual respect. Language barriers cause immediate difficulties for the MNF commander in interoperability with partners.

> *It's difficult to sustain a rapid decision cycle in combined operations.... Even the most common tasks, such as sharing intelligence, must await translation before data can be passed through the command. This, in turn, slows the development of a plan, which, in turn, slows the other elements of the decision cycle.*
>
> General Robert W. Riscassi

Forces must be able to effectively exchange commands, ideas, and information if they are to operate successfully together. Few linguists have both the technical expertise and depth of understanding to cross both language and doctrinal boundaries and be fully understood.

Desert Storm operations provide examples of how language problems can be overcome. The use of US special operations forces (SOF), other personnel with requisite language training, and contracted Kuwaiti nationals provided liaison with Arabic-speaking coalition units. Headquarters, Joint Forces Command placed these teams down to battalion level in the Saudi Arabian, Egyptian, Syrian, and Kuwaiti Armies. In addition to these vertical control mechanisms, specially equipped US Army and Air Force personnel provided horizontal communications between adjacent units to overcome the problem of dissimilar communications equipment. This provided coalition members access to artillery, long-range rocket systems, and tactical and theater strategic air power.

In future multinational operations, differences in language, communications equipment, and procedures will continue. Planners must consider liaison officers, foreign area officers (FAOs), and language-capable personnel to overcome these difficulties. Planners must determine requirements for language-trained personnel early in the planning cycle. These language-qualified personnel will probably require a training period to familiarize themselves with technical terms and procedures of the organization. Linguistic requirements are not confined to liaison officer (LNO) teams or headquarters elements. Linguists are needed throughout the logistics functions as well to coordinate with local authorities, civilian transportation coordinators, refugee and relief centers, hospital staffs, legal offices, and local police forces.

Communications are one of the most difficult challenges in exercising effective C^2 of an MNF. In addition to problems of compatibility and security, many units do not have enough communications equipment to meet mission requirements. Communication planners must play an active role in the MNF's initial operations planning process to identify the required communications architecture to interconnect the MNF. Liaison teams, with adequate communications gear, can reduce the severity of some of these problems. Satellite communications are needed to provide communications between the higher level MNF headquarters to facilitate C^2. Other space-based services, such as weather reporting and use of global positioning systems, may also be needed by elements of the MNF. Communications planners must anticipate these requirements during initial deployment planning, evaluate HN communications resources, and integrate them into the overall communications structure. Continual centralized interfacing between key communications planners during planning, rehearsal, and operational phases helps alleviate interoperability issues. Planners should address issues of spectrum management and controls on access to information systems early in planning.

Chapter 2

Command, Control, Coordination, and Liaison

I was determined from the first to do all in my power to make this a truly Allied force, with real unity of command and centralization of administrative responsibility. Alliances in the past have done no more than name a common foe and 'unity of command' has been a pious aspiration thinly disguising the national prejudices, ambitions and recriminations of high ranking officers, unwilling to subordinate themselves or their forces to a commander of different nationality or different service.

General Dwight D. Eisenhower

The basic purpose of an MNF command is to direct the military effort to reach a common objective. This chapter discusses command structures that have been proven effective, command authorities, and the roles and responsibilities of participants. It also discusses rationalization, standardization, and interoperability; security assistance; information operations; coordination; and liaison.

Multinational operations are unique. Each national commander is responsible to the commander of the MNF, to his national chain of command and, ultimately, for accomplishing his mission. Units maintain a direct line of communications (LOC) to an appropriate national headquarters and thus to their own National Command Authorities (NCA) equivalent. See Figure 2-1.

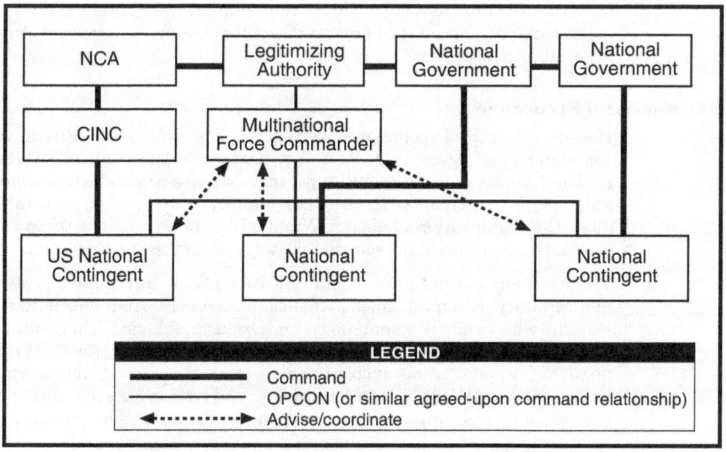

Figure 2-1. Generic Multinational Command Structure

COMMAND STRUCTURE

Multinational operations are categorized in one of two major groups: coalitions or alliances. Coalitions and alliances create a structure that meets the needs, diplomatic realities, constraints, and objectives of the participating nations. Since no single command structure fits the needs of all alliances and coalitions, several different models could evolve. Parallel and lead-nation command structures are discussed under coalitions with integrated command structure presented under alliances. This is normally the case, but not necessarily true for all situations.

COALITIONS

Coalitions normally form as a rapid response to unforeseen crises. During the early stages of such a contingency, nations rely upon their own military command systems to control the activities of their forces. Hence, the initial coalition arrangement most likely involves a parallel command structure. As the coalition matures, the members will probably opt to centralize their efforts through the establishment of a command structure. Some nations call this a *framework* nation. Only the name is different, and since lead nation is better known and more commonly used, all further discussion uses lead nation.

If nations are very similar in cultures, doctrine, training, and equipment, or if extensive cooperative experience exists, an integrated command structure may be effective. This direct approach requires each armed force to receive, understand, plan, and implement missions in the same manner as the other nations. However, C^2 of multinational operations compels commanders to accommodate differences in staff planning capabilities. Some armies have large staffs and the technical means to support planning. Others have austere staff structures and do not have the means to process, reproduce, or rapidly disseminate many contingency plans (CONPLAN). Moreover, decision authority of staffs and subordinate commanders varies among armies. These factors shape the type of command structure chosen. At Appendix A is a guide for coalition operations.

Parallel Command Structure

Under a parallel command structure, no single multinational army commander is designated. Member nations retain control of their own national forces, and the coalition partners write a plan effecting coordination among the participants. Parallel command is the simplest to establish and often the organization of choice. While other command structures emerge as the coalition matures, the parallel model is often the starting point.

The coalition coordination, communications, and integration center (C^3IC) makes unity of effort among coalition members from dissimilar nations possible when unity of command has not been established. This concept proved effective during Operations Desert Shield/Storm. Initially, the C^3IC is the focal point for support issues such as force sustainment, alert and warning, host nation support (HNS), movement control, and training. As a coalition matures, the role of the coordination center expands to include C^2 activities.

When a C^3IC is activated, member nations provide a staff element to the center comprised of action officers who are familiar with support activities such as those discussed above. Coalition nations augment this staff with linguists and requisite communications capabilities to maintain contact with their parent headquarters.

A parallel command consists of two or more headquarters with each having subordinate MNFs, such as during the 1991 Gulf War. Western coalition forces came under control of the US headquarters (thus an example of a command), and Arab forces were under control of the Saudi Arabian headquarters. A graphic depiction of such a parallel command is at Figure 2-2 which shows multinational coalition forces under the control of a dual headquarters. However, the model can be expanded to include additional controlling headquarters as the situation dictates. The C^3IC provides commanders of the parallel commands the capability to bring representatives from all allied armies together quickly to work mutual support issues.

Participating commanders must agree to an internal C^2 structure to oversee operations of the coordination center. During Operation Desert Shield/Storm, only US and Saudi (joint service) representatives staffed the coordination center. Other coalition forces kept abreast of coordination center activities by using LNOs. This proved effective for their situation. The most effective and appropriate way is for other coalition members to provide their own staffs to the coordination center.

US commanders should advocate creation of a similar coordination center in the early stages of any coalition effort operating under a parallel command structure. It is a proven and effective means of enhancing stability and interaction within the coalition as capabilities develop within the theater.

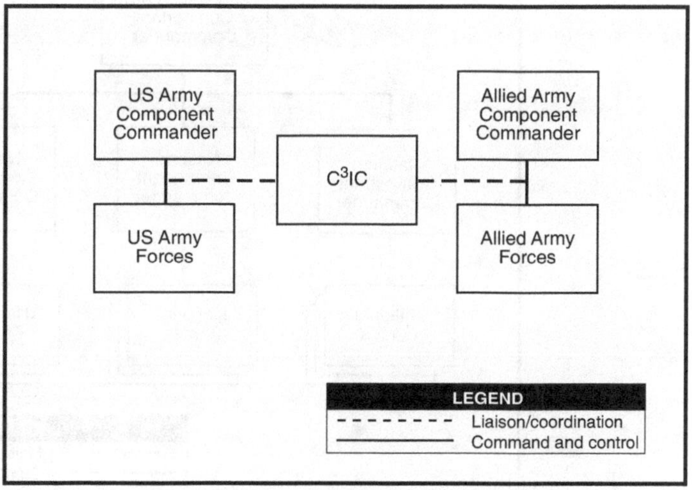

Figure 2-2. Coalition Parallel Command Structure

Lead-Nation Command Structure

The command structure concept recognizes that one nation is assigned the lead role and its C^2 system predominates. Normally, the lead nation provides the largest amount of forces for the operation.

Other nations participating in the operation provide liaison personnel to the headquarters. The commander, working in close consultation with the commanders of the other national contingents, determines appropriate command, control, communications, and intelligence procedures. Robust liaison is essential to developing and maintaining unity of effort in multinational operations. Depending on the size, complexity, and duration of the operation, personnel from the other national contingents can augment staffs. Staff augmentation may also be required if a partner has unique organizations or capabilities not found in forces of the lead nation. This augmentation provides the commander with a ready source of expertise on the respective partners' capabilities during planning and operations.

The command structure is the preferred C^2 arrangement because it achieves unity of command (effort) and supports a more rapid operational tempo (see Figure 2-3). Unique circumstances are required in using the command structure because nations are reluctant to grant this degree of control to another nation. A command structure also presents special challenges for that nation's commanders. Allied counterparts will be particularly sensitive to actions that might be construed as preferential to the lead nation's interests.

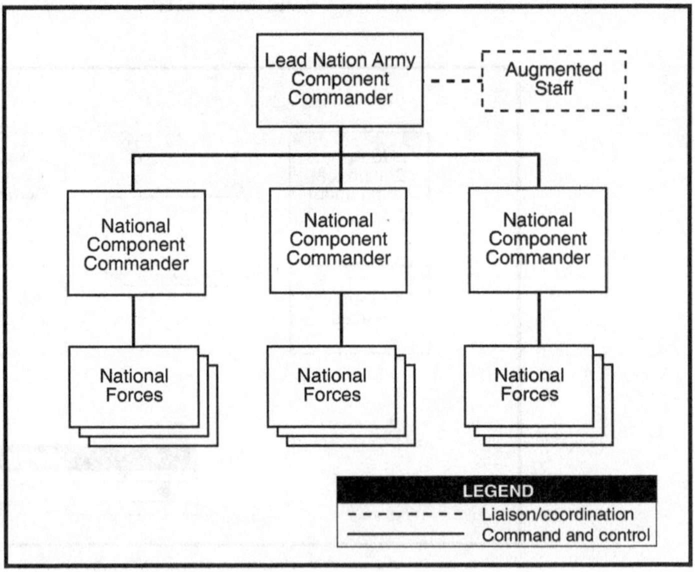

Figure 2-3. Command Structure

ALLIANCES

In an alliance, instead of merely augmenting the staff with other national representatives, the entire staff is integrated. Each primary staff officer could be of a different nationality, and a deputy commander would usually represent the other major participants besides the lead nation. An integrated staff demonstrates greater burden-sharing and commitment, but may create more friction than an augmented staff.

An alliance organized under a multinational integrated command structure provides unity of command in a multinational setting. The NATO command structure is a good example. NATO has a Supreme Allied Commander Europe (SACEUR), who is designated from a member nation. His staff and the commands and staffs of subordinate commands several tiers removed are of multinational makeup.

In alliances, multinational commands are characterized by integrated multiservice staffs at echelons above corps (EAC), with national armies responsive to the multinational commander. This, however, does not negate the possibility of a multinational command and staff arrangement at corps and below. The key ingredients in an integrated alliance command are a single designated commander, a staff composed of representatives from all member nations, and subordinate commands and staffs integrated to the lowest echelon necessary to accomplish the mission (see Figure 2-4).

NATO's Allied Command Europe (ACE) Rapid Reaction Corps (ARRC) is an existing multinational unit. It has some characteristics of an organization in that the United Kingdom (UK) provides most of the framework organization and part of the standing portion of the force. It is, however, an integrated unit, because the primary staff members are of different nationalities. Staffs are integrated (the ARRC G3 is a US brigadier general) and so are two subordinate units, the multinational division (central) and the multinational division

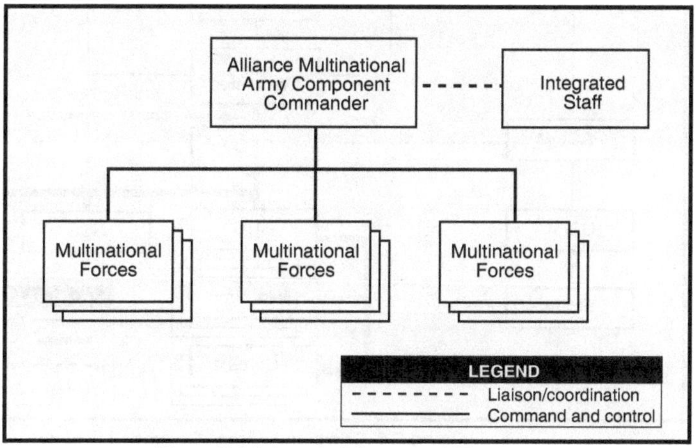

Figure 2-4. Integrated Command Structure

(south). Other NATO nations provide forces for specific operations as the mission dictates. One US division has a habitual relationship of coordinating with the ARRC in peacetime for planning and training. The division is how the US fundamentally forms its warfighting capability. Figure 2-5 shows the complexity of the ARRC organization.

CFC in Korea is an example of a formal multinational headquarters arrangement that has evolved over years of allied cooperation. The CFC staff and the staffs of its subordinate army headquarters (Ground Component Command [GCC]) are fully integrated (see Figure 2-6).

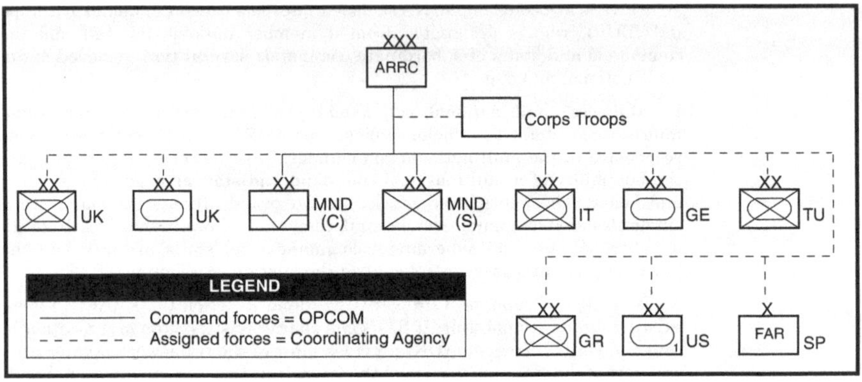

Figure 2-5. ACE Rapid Reaction Corps (ARRC)

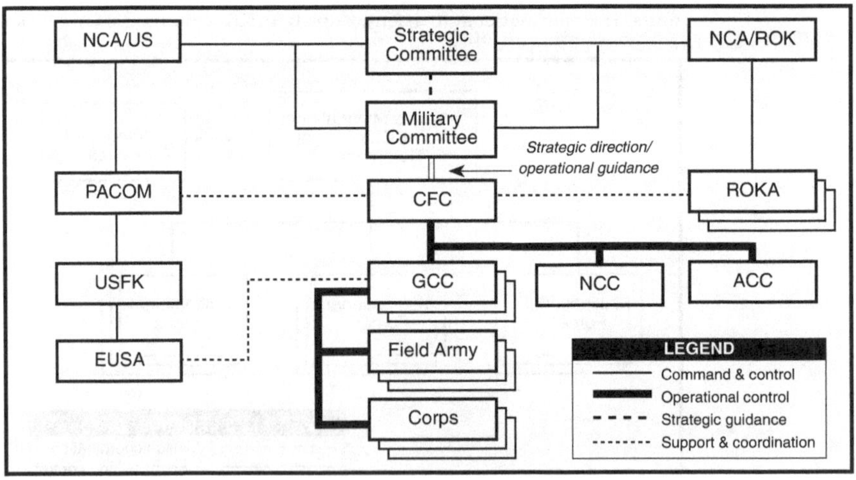

Figure 2-6. Combined Forces Command (Korea)

COMMAND AUTHORITY

When participating in a multinational operation, the senior commanders must agree early on the type of C^2 authority that will govern the operations of the forces. In any multinational operation, the US commander retains command over all assigned US forces. The US chain of command runs from the NCA to the combatant commander. The chain of command, from the President to the lowest US commander in the field, remains inviolate. The definitions shown in this section demonstrate the complexity of multinational operations. Subtle differences in terms, especially *operational control*, cause confusion even among allies with a long history of multinational operations.

Although political considerations are critical, *a clear point must be established where political structure ends and military structure begins.* The MNF commander should report to the combatant commander or a subordinate joint force commander (JFC), who acts as a buffer between political leadership and military structures. This might mean that a US corps commander designated as the commander of the joint task force (CJTF) is the political-military buffer, and the deputy corps commander controls military operations as a joint force land component commander (JFLCC). The combatant commander determines the specific relationship.

US-ONLY OPERATIONS

Authority vested in a commander must be commensurate with the responsibility assigned. In US-only operations, US military forces operate under one of four command relationships:

- Combatant command (COCOM).
- Operational control (OPCON).
- Tactical control (TACON).
- Support (not discussed here).

While US relationships are well defined in joint and service doctrine, they are not necessarily part of the doctrinal lexicon of other nations. NATO and CFC have developed their own terminology for command relationships that meet the needs of those alliances.

The US has approved the definitions of those alliance command relationships. Since the US Army in the future will be primarily a force-projection army, most units will find themselves under one of these relationships; a clear understanding of each is necessary. These definitions provide a framework for forces operating in a coalition force as well. Figure 2-7 depicts the authority inherent in each command relationship.

Combatant Command

COCOM is the command authority authorized by Title 10, *United States Code*, Section 164, or as directed by the President in *The Unified Command Plan* (UCP) to combatant commanders (unified or specified). COCOM provides full authority to the combatant commander to organize and employ commands

Authority	US COCOM	US OPCON	NATO OPCOM	NATO OPCON	CFC/USFK COMBINED OPCON	NATO TACOM	US & NATO TACON
	Most control						Least control
Direct authority to deal with DOD, US diplomatic missions, agencies	X						
Coordinate CINC boundary	X						
Granted to a command	X		X				
Delegated to a command		X		X	X	X	X
Set chain of command to forces	X	X					
Assign mission/designate objective	X	X	X				
Assign tasks	X	X	X			X	
Direct/employ forces	X	X	X	X	X		
Establish maneuver control measures	X	X	X	X	X	X	X
Reassign forces	X						
Retain OPCON	X	X	X				
Delegate OPCON	X	X	X	X with approval			
Assign TACOM	X	X					
Delegate TACON	X	X	X	X	X		
Retain TACON	X	X	X	X			
Deploy forces (information/within theater)	X	X	X	X			
Local direction/control designated forces	X	X					X
Assign separate employment of unit components	X	X	X				
Directive authority for logistics	X						
Direct joint training	X	X					
Exercise command of US forces in MNF	X	X					
Assign/reassign subordinate commanders/officers	X	May suspend or recommend reassignment					
Conduct internal discipline/training	X						

NATO *Full Command* and CFC/USFK *Command less OPCON* are basically equivalent to US COCOM, but only for internal matters

X – has this authority

(shaded box) – denied this authority, or not specifically granted it

LEGEND

COCOM – Combatant command
OPCON – Operational control
OPCOM – Operational command
TACOM – Tactical command
TACON – Tactical control

Figure 2-7. Comparison of Command Authority

and forces as he considers necessary to accomplish assigned missions. This authority enables the commander in chief (CINC) to organize and employ his commands and forces, assign tasks, designate objectives, and give authoritative direction over all aspects of military operations, joint training, and logistics necessary to accomplish the assigned missions. The CINC normally exercises COCOM through his JFCs and service and/or functional component commanders. COCOM is not transferable. Thus, US commanders should anticipate either operating under OPCON of an MFC or being the overall commander of that force.

The combatant commander exercises COCOM authority through one of a combination of six organizational options. These include—

- Service components.
- Functional components.
- Subordinate unified commands.
- Single-service forces.
- Joint task forces (JTFs).
- Direct control over specified operational forces.

Operational Control

Commanders at or below the combatant commander exercise OPCON as their command authority. OPCON is inherent in COCOM and is the authority to perform the function of command over subordinate forces. The CINC may delegate OPCON to his subordinates. OPCON is the most authority with which subordinates can direct all aspects of military operations and joint training needed to accomplish any assigned mission. A commander with OPCON may control forces from one or more services. OPCON does not normally include the authority to direct logistics, administration, discipline, internal organization, or unit training. It does include the authority to prescribe the chain of command and organize commands and forces.

Subject to prior NCA approval, an MNF commander may exercise appropriate and negotiated OPCON over US units in specific operations authorized by a legitimizing authority. One such legitimizing authority is the UN Security Council. In other circumstances, another regional body could be the legitimizing authority. In addition to these control considerations, support relationships and arrangements may often be more appropriate.

Careful judgment of each situation guides actions. Generally, however, the US Army preference is to send a fully capable combined arms force. This force is employed as such under the terms of reference established. If a non-US portion of the MNF requires a certain US-unique capability, the US commander should receive a mission order to support another element for a specific mission and/or operation. Allowing the US to plan and execute the mission reduces friction and simplifies C^2 arrangements.

The degree of OPCON exercised over US units must be coordinated and agreed to between the superior MNF commander and the US-theater CINC who provides the US forces. This agreement must be in consonance with the NCA criteria for the operation's C^2 arrangements. These criteria establish limits to the OPCON that may be exercised over US units. Within these limits, and with few exceptions, a foreign commander cannot—

- Change the mission or deploy US forces outside the area of responsibility (AOR) agreed to by the NCA.
- Separate units, redirect logistics and supplies, administer discipline, promote individuals, or modify the internal organization of US units.

The US commander is ultimately responsible to the MNF commander and to his national chain of command. US units maintain a direct LOC to an appropriate US headquarters—normally the theater combatant commander. Other participants in a coalition maintain similar LOCs.

Tactical Control

The CINC uses TACON to limit the authority to direct the tactical use of combat forces. TACON is authority normally limited to the detailed and specified local direction of movement and maneuver of the tactical force to accomplish an assigned task. TACON does not provide organizational authority or administrative and support responsibilities. The US service component continues to exercise these authorities. TACON differs from TACOM in that TACON involves only the necessary control of movements and maneuvers to accomplish a previously assigned mission.

NATO Operations

NATO control of multinational operations is categorized into five groups—

- Full command.
- Operational command (OPCOM).
- OPCON.
- Tactical command (TACOM).
- TACON.

These NATO definitions do not differ greatly from established US definitions of operational control measures. NATO, using the *standardization agreement* (STANAG) process, has developed, staffed, and gained NATO approval of these definitions, which ensures unambiguous use of these measures, ensuring operational interoperability.

Full Command (NATO)

The NATO equivalent of US COCOM is full command. It is the military authority and responsibility of a superior officer to issue orders to subordinates. It covers every aspect of military operations and administration and exists only within national services. It follows that no NATO or other MNF commander will have full command over other national forces.

Operational Command (NATO OPCOM)

In NATO, OPCOM is the authority granted to a commander—

- To assign missions or tasks to subordinate commanders.
- To deploy units.
- To reassign forces.
- To retain or delegate OPCON and/or TACON as deemed necessary.

OPCOM does not include responsibility for administration or logistics and may denote the forces assigned to a commander. OPCOM allows a commander to specify missions and tasks, assign separate employment to components of assigned units, and reassign forces away from his own force. It does not carry the authority to disrupt the basic organization of a unit to the extent that it cannot readily be given a new task or be redeployed elsewhere. In this area, NATO OPCOM coincides with US OPCON and its authority to *organize and employ commands and forces.* OPCOM allows changing overall organizations and command relationships, but the basic building blocks remain intact. Short of full combat operations, to meet an attack upon NATO territory, US forces will not normally fall under OPCOM of foreign commanders.

Operational Control (NATO OPCON)

OPCON is a defined NATO term. In NATO, OPCON is the authority delegated to a commander to direct forces assigned so that the commander may accomplish specific missions or tasks that are usually limited by function, time, or location. It further includes the deployment of units concerned and the retention or delegation of TACON to those units. Neither does it, of itself, include administrative or logistical control. OPCON is more limited than OPCOM. OPCON does not include the authority to reassign forces or employ a formation, or any part of it, other than on the assigned task, or to disrupt its basic organization so that it cannot readily be given a new task or redeployed elsewhere. Commanders must exercise caution not to interchange US and NATO terms.

Tactical Command (NATO TACOM)

TACOM is defined as the authority delegated to a commander to assign tasks to forces under his command to accomplish the mission assigned by higher authority. TACOM is narrower in application than OPCOM but includes the authority to delegate or retain TACON.

Tactical Control (NATO TACON)

Both NATO and US joint doctrine share the same definition for TACON. The CINC uses TACON to limit the authority to direct the tactical use of combat forces. TACON is the authority normally limited to the detailed and specified local direction of movement and maneuver of the tactical force to accomplish an assigned task. TACON does not provide organizational authority or administrative and support responsibilities. The US service component continues to exercise these authorities. TACON differs from TACOM in that TACON involves only the necessary control of movements and maneuvers to accomplish a previously assigned mission.

CFC/USFK Operations

Combined operations currently employed in the Korean theater require two specific control measures—

- Combined OPCON.
- Command less OPCON.

Although these measures are unique to Korea, their utility is not limited to that theater and should be considered in future OPCON architectures.

Combined OPCON (CFC/USFK)

Combined OPCON does not include the authority to organize assigned and attached forces as necessary to carry out the mission, but strictly refers to employment of warfighting missions. OPCON within CFC is normally limited by function, time, or location. Authority to deploy units concerned and to retain TACON or assign it to a subordinate command is included. Combined OPCON is a more restrictive term than US OPCON in that it includes only the combined warfighting portion of US OPCON.

Command Less OPCON (CFC/USFK)

In Korea, the equivalent to US COCOM is *command less OPCON.* It indicates the national responsibility for all forces and is exercised through appropriate national component commanders. Only warfighting functions are passed to the CINC.

ROLES AND RESPONSIBILITIES

Operations in a multinational environment are both similar to yet different from joint operations. Authority emanates from the NCA to the combatant commanders. Other key players include the MNF commander, the JTF commander, and interagency organizations.

National Command Authorities

Constitutionally, the ultimate authority and responsibility for the national defense rest with the President. The NCA—defined as the President and the Secretary of Defense—may employ military power to respond to situations affecting vital US interests. By law, only the NCA has the authority to direct both the movement of troops and the initiation of military action. COCOM then passes from the NCA to the combatant commanders.

Combatant Commander

The combatant commander translates national strategic direction into theater strategic direction and operational-level objectives for subordinate commanders and exercises COCOM through that chain of command. A combatant commander is normally referred to as the CINC. If he is one of the five CINCs to have a geographic area of responsibility, he is referred to as the theater commander. CINCs develop the theater strategy and campaign plan, organize their theaters, and establish command relationships for effective multinational and joint operations.

Multinational Force Commander

Multinational force commander (MFC) is a general term applied to a commander who exercises command authority over a military force composed of elements from two or more nations. The extent of the MFC's command authority is determined by the participating nations. Such authority, however, is seldom absolute. MFCs unify the effort of the MNF toward common objectives. Gaining consensus is an important aspect of decision making in multinational operations.

Joint Task Force Commander

The CINC usually chooses the JTF organizational option. Thus, the JTF commander may be the senior US commander within an MNF. A JTF is established on a geographical area (examples are Rwanda and Somalia) or a functional area basis when the mission has a specific, limited objective and does not require overall centralized control of logistics. The CJTF exercises OPCON over assigned forces.

Interagency Organizations

The Department of State (DOS) is organized to provide foreign policy advice to the President, nation-to-nation representation throughout the world, US interdepartmental coordination in the various nations with whom the US has relations, and worldwide information services. Several bureaus within DOS have active duty military officers on their staffs. Examples are the Bureau of Political Military Affairs; Office of International Security and Peacekeeping Operations; and the Bureau of International Organization Affairs, Office of Peacekeeping and Humanitarian Operations. Key planners within DOS include—

- Secretary of State.
- US country team.
- Political advisor.
- NGOs and PVOs.

Secretary of State

The Secretary of State is the principal foreign policy adviser to the President. He is responsible for the overall direction, coordination, and supervision of US foreign relations and for the interdepartmental activities of the US Government overseas.

US Country Team

The US country team is the senior, in country, US coordinating and supervising body. Headed by the chief of the US diplomatic mission, the team is composed of the senior member of each represented US department or agency, as desired by the chief of the US diplomatic mission (Joint Pub 1-02). The team includes representatives of all in-country US Government departments. The ambassador represents the President as chief of mission, but takes policy guidance from the Secretary of State through regional bureaus. The ambassador is responsible for all US activities within the country to which accredited and interprets US policies and strategy regarding the nation.

Composition of the country team varies widely depending on specific US national interests in the country, the desires of the chief of mission, the situation within the country, and the number and level of presence of US agencies. The ambassador's authority does not, however, include the direction of US military forces operating in the field when such forces are under the command of a US area military command. The CINC of the US area military command usually participates as a member of each country team, even though he is not a member of the diplomatic mission and may not be physically located in the country. The country team would have direct impact on military operations through—

- Negotiating status of forces agreements (SOFAs).
- Negotiating basing rights.
- Providing firsthand knowledge and advice to military units.
- Performing other similar actions.

Political Advisor

A political advisor (POLAD) is a foreign service officer from the DOS. The DOS assigns a POLAD to each combatant commander and may authorize one to the American operational commander during operations, such as the Rwanda relief operation Support Hope. The POLAD is a valuable asset with appropriate regional knowledge and skills that can help the combatant commander in translating political objectives into military objectives. Additionally, the POLAD can facilitate cooperation between the primary US political and military actors. The POLAD often is able to move freely throughout an area of operations (AO) and work with a wide range of different parties that might not work with US military personnel.

NGOs and PVOs

The primacy of political considerations in multinational operations demands a recognition of the importance of nonmilitary organizations. NGOs and PVOs are frequently on the scene before military forces and are willing to operate in high-risk areas. They will most likely remain long after military forces have departed. Clear, prearranged relationships with NGOs and PVOs need to be established, especially during short-notice deployments or potentially dynamic scenarios. The sheer number of lives they affect and the resources they provide enable the NGO and PVO community to wield a great deal of power. Because of their capability to respond quickly and effectively to crisis, NGOs and PVOs can lessen the civil-military resources that a commander would otherwise have to devote to an operation.

Finally, activities and capabilities of NGOs and PVOs must be factored into the commander's assessment of conditions and resources and integrated into the selected course of action. NGOs and PVOs play a major role in most multinational military operations, but their role lessens as the military operation gets closer to war. Examples of such organizations include—

- The International Society of the Red Cross and Red Crescent (commonly called the *Red Cross*).
- The UN High Commissioner for Refugees, *Medecines Sans Frontiers.*

HISTORICAL EXAMPLE

US forces were deployed to Central Africa to provide humanitarian assistance in the wake of the ethnic violence in Rwanda and the massive cross-border flight of refugees. Other forces and relief organizations were already responding.The fluid situation and short duration of the operation, along with the urgency of the response and the need for establishing support bases in other countries, did not allow for a formal multinational command structure. Colonel James McDonough, commander of the army force there, described operations this way:

> We cooperate (vice combine) with the forces of other nations (essentially a UN peacekeeping and humanitarian force), the US ambassador's country team, the many NGOs and UN agencies, and the local government officials, private contractors, and entrepreneurs that emerge from the ruin of this country.

RATIONALIZATION, STANDARDIZATION, AND INTEROPERABILITY

Achieving and implementing international rationalization, standardization, and interoperability (RSI) with allies, coalition partners, and other friendly nations is key—

- To achieving the closest practicable cooperation among their military forces.
- To achieving the most efficient use of research, development, procurement, support, and production resources.
- To agreeing, where applicable, to emphasize activities and initiatives that provide the means for a more effective multinational warfighting capability.

International military RSI applies to both materiel and nonmateriel matters.

RATIONALIZATION

Rationalization is defined as any action that increases the effectiveness of allied forces through a more efficient or effective use of defense resources committed to the alliance. Rationalization includes consolidation, reassignment of national priorities to higher alliance needs, standardization, specialization, mutual support or improved interoperability, and greater cooperation. Rationalization applies to both weapons and materiel resources and nonweapons military matters.

STANDARDIZATION

Unity of effort is greatly enhanced through *standardization*. The basic purpose of standardization programs is to achieve the closest practical cooperation among MNFs. This is achieved through the efficient use of resources and the reduction of operational, logistical, technical, and procedural obstacles in multinational military operations.

In established alliances, armies usually develop a degree of standardization regarding administrative, logistical, and operational procedures. The mechanisms are STANAGs, and they are binding on all US forces.

Partners of a long-standing alliance are usually familiar with one another's equipment and have established some degree of interoperability. STANAGs are the instruments in NATO for establishing commonality in procedure and equipment. The *quadripartite standardization agreements* (QSTAGs), agreed to within the ABCA armies program (and usually identical to STANAGs), are another excellent example of these types of agreements.

Both STANAGs and QSTAGs provide a baseline for cooperation within a coalition. In many parts of the world, other bilateral agreements for interoperability among potential coalition members may already be in place prior to the formation of the coalition. However, in most *ad hoc* coalitions, not all participants are immediately familiar with such agreements. The multinational commander must initially rely on designated unit standing operating procedures (SOPs) and clearly written, uncomplicated operations orders.

Implementation of STANAGs, either NATO or ABCA, is transparent to US units. These mechanisms obviate the need for alliance-unique doctrine. The soldiers of each alliance participant use national doctrine/TTP, and, to them, the interoperability is transparent. Implementation occurs when the content of the STANAG is incorporated into Army doctrine. For example, within NATO, operational terms and symbols are agreed upon, and each nation, to a high degree, ensures their internal symbology conforms to the standard. This way, US Army forces use the map symbols and terms in FM 101-5-1. Except for some country-unique symbols, a new set of symbols to conduct operations with NATO allies is not needed. The five-paragraph operations order, liaison procedures, air support procedures, engineer target folders, and letters for designating days and hours of an operation are examples of existing agreements.

INTEROPERABILITY

Interoperability is defined as the ability of systems, units, or forces to provide services and to accept services from other services, units, or forces. Use of the exchanged services enables them to operate effectively together. Historically, the problems of interoperability have been solved primarily through trial and error during actual conduct of operations over an extended period of time.

Factors that enhance interoperability start with adhering to the principles, considerations, and tenets of multinational operations. Additional factors include—

- Planning for interoperability.
- Knowing personalities of the commander and staff.
- Assessing allied capabilities.
- Promoting a command atmosphere that permits positive criticism.
- Providing liaison teams.
- Constantly striving to eliminate sources of confusion and misunderstanding.

Factors that restrict interoperability include time; refusal to cooperate with partners; level of differences in military organization, doctrine, equipment, and experience; and conflicting personalities.

SECURITY ASSISTANCE

The Army Security Assistance (SA) Program is an essential complement to the overall US defense effort. SA covers the transfer of military and economic assistance through sale, grant, lease, or loan to friendly foreign governments. Transfers are carried out under the principle that if they are essential to the security and economic well-being of governments and international organizations, they are equally vital to the security and well-being of the US.

US national security interests, increased regional tensions or threats, or actual conflict create a need for SA. To increase readiness, assets in this program can be diverted from a customer to either US or allied forces in direct support of theater requirements.

Planners must determine what SA has been or is being provided to members of the MNF, its impact on commonality of equipment, and the degree of adoption of US Army doctrine and tactics in the use of SA. This commonality can be the foundation to develop cooperation and interoperability among members of the MNF.

Material and services committed to SA requirements cannot be reallocated to US forces without Secretary of Defense approval. The US Army ensures that auxiliary equipment, logistics support, and training are available before a major end item of SA equipment is provided to a friendly foreign nation.

The Defense Security Assistance Agency (DSAA) directs, administers, and supervises the execution of SA programs through CINC SA divisions. This involves assisting military services, unified commands, and in-country SA officers in their efforts to help foreign governments obtain US equipment, training, and other defense-related services authorized by the *Foreign Assistance Act* (FAA), as amended, and the *Arms Export Control Act*. SOF units are well suited to conduct SA because of regional orientation, language capability, and the requirement to train foreign forces. Recently, DSAA assisted US Atlantic Command (USACOM) in getting equipment for the multinational military force that supported the US forces in Operation Uphold Democracy.

INFORMATION OPERATIONS

Information operations are continuous military operations within the military information environment (MIE). Information operations enable, enhance, and protect the commander's decision cycle and mission execution to achieve an information advantage across the full range of military operations. They include interacting with the global information environment (GIE) and exploiting or denying an adversary's information and decision systems. FM 100-6 provides an in-depth breakdown of US Army information operations.

Information operations integrate all aspects of information to support and enhance the elements of combat power. The goal of information operations is to dominate the battlespace at the right time, right place, and with the right weapons or resources. Units conduct information operations across the full range of military operations, from operations in garrison, through deployment, to combat, to redeployment. Information operations apply to information and the systems that produce, acquire, process, store, or disseminate information. Activities that support information operations include acquiring, using, protecting, managing, exploiting, and denying information and information systems. These operations take place within three interrelated components of information operations: operations (command and control warfare [C^2W], civil affairs [CA], and public affairs [PA]), relevant information and intelligence (RII), and information systems (INFOSYS). These activities operate within a layered battlespace comprised of the GIE and MIE. Army organizations conduct these information operations activities as part of a dynamic, interactive process in support of each component in an integrated full-dimensional operation.

Information operations are inherently multiservice and multinational. The development of information operations capabilities, plans, programs, tactics, employment concepts, intelligence, and communications support as a part of military strategy requires close coordination with responsible Department of Defense (DOD) components and partners in the alliance or coalition. Key to information operations in multinational operations is the need to plan in a multinational manner and achieve a workable multilevel security program. The J2 achieves this. An exchange of LNOs may be the most effective way to secure these objectives.

Information operations must carefully account for how the GIE affects operations. Public perception can put political pressure on nations to modify their participation in the effort. Many countries and their political decision makers do not get the same amount of information; they do not have the same communications technology that is available to the US. As a result, their frame of reference may be what they see on television. Information operations must expand the MNF frame of reference both technologically and intellectually.

US forces are familiar with US media organizations and their methods, but other national media elements operate under different rules. Public affairs officers and commanders must realize the differences and work closely with the media to develop an open environment with a minimum of ground rules to maintain operational security.

COORDINATION

In multinational operations, coordination is critical among both the MNF and interagency organizations. Almost all operations involve NGOs and PVOs. Given these circumstances, coordination centers are the key to getting things done. A commander may establish any type of coordination center he chooses.

MILITARY COORDINATION CENTER

This type of coordination center is used where no real command structure effectively exists. All MNF participants should be represented and attend the daily coordination meeting. This is not a preferred structure, but it may be the only feasible option. In Operation Provide Comfort, this center included Iraqis in the daily meetings. This ensured clarity of communications and avoided potential confrontations by providing a forum to negotiate and find acceptable ways to accomplish the mission.

CIVIL-MILITARY OPERATIONS CENTER

The civil-military operations center (CMOC) is the coordination center established and tailored to assist the commander's civil-military operations (CMO) officer. The CMOC assists in anticipating, facilitating, coordinating, and orchestrating those civil-military functions and activities pertaining to the civilian population, government, and economy of areas in which armed forces, government organizations (GOs), international organizations (IOs), NGOs, and PVOs are employed. The CMOC is neither a unit nor an organization; it is simply a capability or extension provided by the CMO officer that facilitates access to civilian agencies and nonmilitary organizations participating in or having peripheral interest in a particular operation. As an extension or capability of the force CMO officer, the CMOC reports and transmits data (generally in the form of request for assistance) for the GO, IO, NGO, and PVO representatives directly to the force officer.

The CMOC coordinates and facilitates US and MNF operations with those of GOs, IOs, NGOs, and PVOs, HN agencies, and HN authorities. The CMOC provides access for nonmilitary agencies desiring assistance and coordination from the military. It serves as an extension of the civil-military cell providing both access and coordinating-authority-related data or information from and to nonmilitary agencies operating away from the military headquarters.

The CMOC may be comprised of, or may be augmented by, military and/or civilian representatives from many different agencies. Mission requirements, command directives, operational security, work load, and accessibility to nonmilitary agencies impact on the actual organization of the CMOC.

The number of CMOCs supporting a given operation may vary based on mission analysis and distance from the headquarters serving a particular geographic or tactical area. In operations where the JFC's headquarters and the majority of subordinate units are located in proximity to the civilian/HN diplomatic center and GO, IO, NGO, and PVO representatives, a CMOC may be established to facilitate access by those organizations. Conversely, in operations where the joint force headquarters is located in one locale and subordinate units are spread throughout the country, subordinate commanders may establish sector CMOCs to provide the same type of civil-military facilitation. In addition to sector CMOCs, military commanders may also have to establish CMOCs at every level of command from unified command down to battalion level. Again, this would depend on the geographic area and tactical control measures.

COALITION SUPPORT TEAMS

Coalition support improves the interaction of coalition partners and US military forces. It includes—

- Training coalition partners on tactics and techniques.
- Assisting with communications interfaces to integrate them into the coalition command and intelligence structure.
- Establishing liaison to coordinate combat support and combat service support (CSS).

Coalition support teams often provide the JFC with an accurate evaluation of the capabilities, location, and activities of coalition forces, thus facilitating JFC C^2. A coalition support team was first employed during Operations Desert Shield/Storm and has been critical in subsequent multinational operations. Past success in these operations in assisting and integrating coalition units into multinational military operations has made it likely that coalition support will become an important aspect of future multinational operations.

US Special Forces can serve as coalition support teams. Coalition support teams serve to build cohesion and synchronize operations; establish a parallel coalition command, control, communications, and intelligence (C^3I) system; provide on-the-ground information; and coordinate air support. Special Forces units are well suited to accomplish these missions because of their regional orientation, language capability, and requirement to train foreign forces.

LIAISON

Use of LNOs is essential to the success of MNF operations. The multinational commander has neither the time nor sufficient interoperable C^2 systems to establish effective C^2 without the use of LNOs. Even in alliances, where armies often have long established interactions and a high degree of standardization in equipment and procedures, strong reliance is still placed on LNOs. In Korea, with the automated theater bilingual C^2 system, TACCIMS (Theater Automated Command and Control Information Management System), heavy emphasis is on liaison teams. If possible, LNOs coordinating with a country should speak the language. However, the professional abilities of the LNO are the determining factor. LNOs not conversant in the necessary foreign language require interpreters.

Liaison requirements must be identified early in the planning process, and exchange of personnel must be reciprocal. Any changes in task organization may create additional requirements. LNOs follow the normal pattern of higher to lower, left to right, and supporting to supported and must be able to speak for their commanders. As representatives of their commanders, they attend briefings and maintain close contact with the MNF staff, especially the operations staff. For specific operations or when one nation supports another, they exchange specialists such as aviation staff officers, fire support officers, engineers, or intelligence specialists. CA, psychological operations (PSYOP) staff officers, and legal advisors are necessary to deal with HN and interagency matters. See Appendix B for additional information.

HISTORICAL EXAMPLE

During Desert Shield/Storm, one major tactical concern was the ability of US and French forces to fight effectively side by side with no previous experience in doing so. Communicating each other's intentions and passing that information rapidly up and down the bilingual chain of command were challenges to be met. The first step to allay these concerns was to exchange bilingual liaison teams at brigade and higher headquarters.

XVIII Airborne Corps headquarters, 82d Airborne Division, 24th Infantry Division (Mechanized), 101st Air Assault Division, and the 18th Field Artillery Brigade all exchanged liaison teams with the 6th Light Armor Division. These teams had a primary mission to ensure the passage of accurate and timely information. They accomplished this by using organic US radio equipment between the Daguet Division headquarters and their parent unit's headquarters. The teams also served as sources of information on the doctrine, tactics, SOPs, force structure, and capabilities of their respective units. Over time, these liaison teams were integrated into the tactical planning process at Daguet Division headquarters and educated the staff sections to which they were assigned on the finer details of their unit's capabilities. This same process was occurring at XVIII Airborne Corps and subordinate US headquarters, where bilingual French liaison teams were involved in planning at every headquarters to which they were assigned.

To ensure accurate and timely indirect fire during the operation, a US Army fire control system (TACFIRE) detachment was integrated into the French fire support coordination center at Daguet Division headquarters to orchestrate fire coordination measures. This ensured face-to-face coordination between US and French artillerymen at the decision-making point.

The LNO must be ruthless in his quest for information and, while observing protocol, must attempt to accompany the American commander to as many high level meetings as possible. At this stage, the LNO can get inside the American' thought process, which is invaluable to his own commander.

Translated from another nation's exercise AAR

Chapter 3

Operational-Level Considerations

This task requires implementation of a carefully thought out, sustained and progressive peacetime process of development, training, and integration in all the critical areas of...operations. Interoperability, mutual confidence, and success cannot be obtained on the brink of a conflict, nor can they be achieved by a sudden and improvised effort. Good intentions cannot replace professional preparations.

Commander Juan Carlos Neves, Argentine Navy

Military capabilities differ based on national interests and objectives, national character, doctrine, training, leader development, organizations, and materiel. Some doctrines emphasize offensive operations while others emphasize defensive operations. Some nations prepare for highly mobile, mechanized operations, while others concern themselves with counterinsurgency operations. The MFC must know and consider these differences when assigning missions and conducting operations. This chapter provides operational-level considerations to improve the effectiveness of US forces when operating with the MNF.

A variety of functions helps commanders build and sustain combat power. In general terms, these are called *combat functions* and are described fully in FM 100-5. Army commanders and staffs translate combat functions into more specific operational-level functions when conducting operational-level missions and tactical-level functions when conducting tactical operations. The five operational-level functions discussed in this chapter are *movement and maneuver*, *intelligence*, *firepower*, *support*, and *protection.*

At the operational level, C^2, intelligence, and support are critical to building an effective MNF. A major weakness in one of these areas is a far greater threat than one in other systems. All other elements hinge on the effective integration of these three systems. Major differences, real or perceived, among alliance or coalition partners cannot be tolerated. Within these operating systems, effective liaison and language capabilities make effective operations possible and alleviate friction and confusion.

MOVEMENT AND MANEUVER

Movement and maneuver involve the disposition of the MNF to create a decisive impact on the conduct of subordinate campaigns or major operations. This is achieved by securing the operational advantage of position before battle is joined or exploiting tactical success to achieve operational success. This operating system includes the movement or deployment of force for operational advantage and the conduct of maneuver to operational depths (for offensive or defensive purposes). It also includes the enhancement of the mobility of friendly forces, to include—

- Ensuring noninterference from dislocated civilians, enemy prisoners of war (EPW), and other civilian internees (CI).
- Degrading the mobility of enemy forces.
- Controlling land, sea, or aerospace for operational advantage.

Movement and maneuver can be on sea and land or through the air.

OPERATIONAL MOVEMENT

Operational movement is critical to MNF operations. An MNF headquarters, or supporting CINC, normally plans and executes all strategic movement, but it remains a national responsibility to move forces into the operating area. The MNF headquarters is responsible for coordinating these strategic deployments to support the commander's plan and then for planning and controlling intratheater movement through the reception, staging, onward movement, and integration (RSO&I) process.

Intermediate Staging Base

The decision on use of an intermediate staging base (ISB) outside the operating area requires a conscious decision to balance numerous requirements, including greater lift requirements against better efficiency. MNFs should assemble and stage in an ISB, especially if combat is imminent. The ISB provides a secure area to assemble, train, equip, and bond the coalition or alliance force into a cohesive one. It is better to solve problems and correct deficiencies in a nonhostile environment. The MNF staff should assemble first and work together, followed by the rest of the MNF force. The ISB should have sufficient billeting and training capacity to support the entire MNF at once. Access to airports and seaports for smooth reception of the force and its subsequent deployment is critical. If the multinational operation is a lesser regional contingency or a second major regional contingency (MRC), an ISB may not be possible. This is because of limited strategic lift (air and sea), size of MNFs, or availability of an HN in theater for use as an ISB.

Planners should anticipate disruptions from many factors, ranging from weather to political decisions, that alter the planned flow of personnel, forces, and equipment. The ASCC, or a supporting CINC, operates the ISB and deploys fully ready forces into the operating area. When the lodgment is well established, the ISB may shift into the operating area if it will not drain additional resources. Larger areas, such as Europe or Korea, allow for a reception center or ISB in theater from the beginning. In the Korean War, UN forces operated the UN reception center where incoming contingents were equipped, trained, and linked up with US liaison elements.

Theater Reception

At ports of debarkation (PODs), units work with HNS and multiservice personnel to secure the POD, discharge equipment, process equipment and personnel, and move units to marshaling areas. HN forces can perform and assist in many of these functions. If no HNS forces are available, national units must perform those functions. All units must be planned for in the strategic flow and some, because of their limited capability, may be assigned a certain function.

Theater Staging

Staging is that part of the RSO&I operation in which several key activities take place in controlled areas in the ongoing incremental buildup of combat power. During this phase, units are reassembled and united with their equipment and scheduled for movement toward the tactical assembly area; materiel is segregated, prioritized, and prepared for transport; class V supplies are uploaded; and life support is provided to personnel.

Theater Onward Movement

This phase begins when units are configured to move to their final destination. Such movement is accomplished through a carefully devised movement program that employs convoy, rail, and HN contract assets (such as heavy equipment transporters and other trucks) to ensure the forward and concurrent movement of troops and supplies. Centralized control of transportation assets is required. Real estate management may be a problem unless a multinational counterpart to the US joint force utilization board (usually controlled by engineers) is given authority to allocate terrain to all forces and agencies.

Movement planning must account for differences in how nations conduct road marches or similar administrative movements. These differences can lead to confusion and disorganization. For example, one MNF contingent may consider any movement made by a tactical unit to be a tactical movement. Terminology must be coordinated through the movement control center to avoid confusion.

When planning the movement of MNFs, planners must know the details of the organization, equipment, capabilities, and limitations of the forces. Planners must know how to efficiently request intratheater movement of multinational operational forces consistent with the operational commander's operations plan (OPLAN). The movement should complement sequencing of operations and time-phased force deployment. Movement planners should consider all assets (joint, allied, HN, and third country) and modes (air, land, or sea) of transport. During execution of these movements, movement control personnel must locate where they can validate actual movements.

Theater Integration

This phase covers both the effective management of reception, staging, and onward movement of units and the transfer of authority (TOA) of units to the tactical commander.

Operational Maneuver

Operations in depth include both conventional forces and SOF. To best achieve operational aims, plans should reflect the special capabilities of each national contingent in the assignment of missions. Significant factors at this level include—

- Mobility, size, and sustainability of formations.
- Training for operations special environments.
- Preparing for operations involving nuclear, biological, and chemical (NBC) weapons.

Differences in MNF doctrine, materiel, and national styles become evident when forces conduct continuous operations, penetrations, encirclements, exploitations, or pursuits.

Operational Mobility and Countermobility

Operational mobility facilitates the movement of the MNF in a campaign or major operation without delays caused by operationally significant terrain, obstacles, or dislocated civilians, EPW, and CI. Operational countermobility delays, channels, and stops the air, land, and sea movement of enemy operational formations. It creates positional advantage for friendly forces and exposes enemy centers of gravity or high-payoff targets for destruction. The principle of centralized planning and decentralized execution applies. Engineers play a key role in mobility and countermobility, and national forces vary greatly in equipment and organization. The MNF commander must ensure missions are assigned appropriately to achieve desired results and avoid unnecessary casualties.

INTELLIGENCE

Operational intelligence is required to plan and conduct MNF operations. Allied and coalition partners normally operate separate intelligence systems to support their own policy and military forces. Intelligence systems supporting MNF operations should focus on collection, analysis, and dissemination of information to identify and locate the threat centers of gravity. They should also identify potential weaknesses or targets within the friendly MNF. Effective MNF intelligence operations—

- Define the end state in terms of information to be shared and work from the beginning of planning to make that happen.
- Define dissemination architecture for all partner forces.
- Agree on and disseminate common maps.

Flexibility is essential to adjust to each nation's expectations of intelligence support. Sharing of intelligence sources and methods, to include cooperative intelligence collection and production, helps coalition members attain common objectives. National systems vary widely in sophistication and focus. Most allies cannot approach the range of US technical capabilities to collect and process intelligence. Yet, many MNF partners have greater ground reconnaissance and human intelligence (HUMINT) capabilities than do US units. Questions on the amount of intelligence a US commander is permitted to share with non-US partners surface frequently.

The MNF commander must ensure rapid dissemination of military intelligence and use of available intelligence assets by all partners. Limits and procedures for sharing intelligence must be confirmed during initial coordination and negotiations among the MNF participants. This usually requires the formation of a multinational intelligence staff at the highest level. Once the senior US commander establishes the procedures and limits for sharing intelligence, the J2 establishes the system to provide the information. In many cases, an intelligence support element (ISE) is provided to the coalition or alliance headquarters. This may require an intelligence network with dedicated communications and LNOs to provide appropriate interface among headquarters.

Prior to receipt of instructions, initial guidance can be obtained from National Disclosure Policy (NDP-1). This policy promulgates national policy and procedures in the form of—

- Specific disclosure criteria and limitations.
- Definitions of terms.
- Release arrangements.
- Other guidance required by US departments and agencies having occasion to release classified US military information to foreign governments and IOs.

In addition, it establishes and provides for the management of interagency mechanisms and procedures required for the effective implementation of the policy. When guidance is lacking, only perishable information that immediately impacts tactical operations may be shared.

FIREPOWER

Operational firepower, according to US Army doctrine, is the application of firepower and nonlethal means to decisively impact the conduct of a campaign or major operation. Operational firepower is a separate component of the operational scheme and the coequal of movement and maneuver, but maneuver and firepower must be fully integrated. Operational firepower is not fire support (close air support [CAS], artillery, mortars, rockets, missiles, etc.), and operational maneuver is not necessarily dependent on operational firepower. However, operational maneuver can be affected by operational firepower. Assets (mostly joint) other than those required for the routine support of tactical maneuver normally furnish operational firepower. They include deep strike assets (ATACMs) and attack helicopters on deep missions. Other national forces do not normally have the deep attack capabilities of US forces. The US is unique in using aviation assets as a maneuver combat force capable of independent offensive action.

Effective C^2 of MNF firepower assets is the key. Operational firepower must be coordinated with the joint force air component commander's (JFACC) air operations plan through a multinational targeting coordination board. The joint targeting coordination board (JTCB), if established, may be subordinate to the multinational board. In Korea, the combined targeting board already exists and works under the JFACC. This is an example of how terminology may differ slightly in different situations, but the function is more important than the name. Most major powers use similar firepower terminology such as *fire support coordination line* (FSCL).

Nonlethal attack intends to impair, disrupt, or delay the performance of enemy (or adversary) forces and their facilities. It includes PSYOP, use of SOF, electronic warfare (EW), and the command, control, communications, and countermeasures subset of information operations. Planners must pay close attention to the effect such operations may have on members of the MNF. Although PSYOPs are directed at adversaries/enemies, some MNF members may have cultural, religious, or ethnic ties to the targets.

Integrated joint and MNFs must achieve a balanced mix of all available joint and allied air defense (AD) forces. They must establish positive control measures to minimize mutual interference or fratricide. Also, easily identifiable means of identification to ensure survival of friendly aircraft if positive control measures fail must be established. Some nations have air defense artillery (ADA) assets with the army while others are with the air force. Some divide ADA assets between both services. Differences in procedures must be understood throughout the MNF.

PSYOPs are planned operations to convey selected information to foreign audiences to influence their emotions, motives, objective reasoning, and, ultimately, the behavior of foreign governments, organizations, groups, and individuals. The purpose of PSYOP is to induce or reinforce foreign attitudes and behaviors favorable to the originator's objectives.

As a force multiplier, PSYOP is one of the most effective weapons available to a commander. It can reduce casualties on both sides by—

- Reducing the morale and combat effectiveness of the enemy.
- Creating dissidence and disaffection within their ranks.
- Encouraging surrenders or defections.
- Promoting resistance or inducing within a civilian populace a lack of support for a hostile regime.

PSYOP also has proven to be effective in peacetime operations, supported SOF, conventional forces, other US Government agencies, and foreign governments.

SUPPORT

Operational support ties requirements to strategic capabilities to accomplish operational plans. It encompasses support required to sustain both the tempo and continuity of joint and multinational campaigns and other military activities within the AOR. The major CSS areas of concern are reception, staging, onward movement, integration, positioning of facilities, materiel management, distribution, reconstitution, and redeployment.

Multinational logistics presents major challenges. Included are differences in doctrine, stockage levels, mobility, interoperability, infrastructure, national resource limitations, and units of measurement. Multinational operational-level logistics focus on how to integrate sustainment operations through coordinating and executing a cohesive logistics program. Traditionally, national differences have made logistics solely a national responsibility.

Future multinational operations will require greater compatibility, and support must be the collective responsibility of the nations involved. For example, NATO's logistics principles give nations and NATO authorities a collective responsibility for logistics support of NATO's multinational operations. The NATO commander at the appropriate level must have sufficient authority over logistics resources to enable him to employ and sustain his forces in the most effective manner. Varying methods of mutual logistical support must complement the partners' capabilities and minimize their weaknesses.

Personnel service support (PSS) includes replacement operations; strength management; personnel accounting; casualty management; personnel data base management; personnel information management; postal operations; enlisted and officer evaluations; promotions; transfers and discharges; morale, welfare, and recreation; and awards. It must be planned and coordinated within personnel channels of US forces and other MNFs.

PLANNING

The level of participation by US Army support forces depends upon the specific political and military objectives agreed to at the national level in consonance with US national military strategy. Lacking existing international agreements, combatant commanders must have NCA to enter into relationships (providing or accepting CSS from multinational partners) that are contrary to US policy. While execution of multinational support requires approval, planning does not. Full coordination with potential allies needs to be aggressively pursued before and during operations. Planning considerations include—

- MNF composition.
- Compatibility.
- HNS limitations.
- Overall infrastructure conditions in the AO (for example, electricity, water, roads, rail, communications).
- Budget limitations.

Any support provided to multinational partners is in addition to established executive agent support responsibilities that the Army has to other US services, civilian contractors, DOD, and other US government agency personnel. It does not take priority unless so directed by the NCA.

The objective is to achieve a realistic degree of compatibility of equipment, supplies, and procedures. This is within the mission, enemy, terrain, troops, and time available (METT-T) constraints, political and legal realities, and the participating nations' capabilities. The intent is to introduce units to support the MNF in achieving the political and military goals with minimal duplication. To do this, leaders and their staffs must understand the organization and capabilities of all national contingents. Such knowledge allows them to compensate for differences and shortfalls during plan formulation and remain flexible during execution. Some major differences between national forces include doctrine, organizational structures, equipment, SOP (to include reporting formats), terminology and definitions, requirements computation methods, sustainment planning criteria, common supply resources, support services, reconstitution policies and procedures, and automated data processing (ADP) support systems. Within NATO and the ABCA armies program, STANAGs and QSTAGs have been developed to assist planners in multinational operations. Nations outside of these programs could possibly use these documents as guidance for support planning.

Combat Service Support Structure

Planners must resolve several key subjects as they build the CSS structure of the MNF. Some common concerns include—

- Identification of common supplies and services that might be provided to all participants by one nation (role specialization).
- Chapter 138, title 10 US Code authorizes the Secretary of Defense to negotiate Mutual Logistics Support Agreements or Acquisition Cross-Service Agreement[1] (ACSA) with friendly nations in which US forces are deployed. ACSA permits the reciprocal transfer of logistics support, supplies, and services between the armed forces of the US and the armed forces of NATO countries and other designated countries outside of NATO including Japan. ACSA permits the multinational commander to acquire and transfer logistics items or services for cash, replacement-in-kind, or equal value exchange.
- Development of the means to maintain national asset visibility (from national sustaining bases to the front-line units).
- Procedures to obtain and update the readiness and stock status of national forces.
- Ensurance of compatibility of communications networks including ADP interfaces among multinational and national support systems.
- Security.

Combat Service Support Template

The CSS structures that support multinational contingency planning include selective functional areas and subareas available to multinational commanders and staffs during planning and execution of operations. Most nations use individually developed operational and logistical employment concepts and terms. The employment concepts and terms used here represent an acceptable standard for contingency operations in most US forces' multinational operations.

Multinational Support Command and Control

Multinational support C^2 is extremely complex and may require a supreme effort to coordinate the collective resources of all assigned forces. The basic concept of C^2 is an integrated system comprised of doctrine, procedures, organizational structure, personnel, equipment, facilities, and communications. This system then provides authorities at all levels with

[1] A formal decision between two or more nations that delineates the procedures to release national assets to forces of the other nations.

timely and adequate data to plan, direct, and control their activities. The CSS staff must maintain close contact with all units, customers, providers, G3 staffs, staffs of higher and lower formations, and, especially, with headquarters of CSS forces under their control.

Multinational Support Command/Element

The multinational support command element (MSC/E) prioritizes and synchronizes key CSS functions to best use the collective assets and resources of all contributing nations. The MSC/E is a generic, CSS headquarters created for the multinational operation. The primary tasks of the MSC/E staff, composed of both the multinational commander's CSS staff and CSS representatives from each contributing nation, are to coordinate the efforts of assigned multinational CSS organizations, to help resolve international and interservice issues, and to coordinate the use of scarce resources such as strategic and tactical lift. Furthermore, the MSC/E sets the sustainment phasing and timing of the operation, as well as the priorities for assets and resources for the MNF. Its primary responsibility is to monitor and access the entire CSS situation; to translate operational guidance, plans, and orders into CSS requirements; to issue directives; and to implement CSS priorities.

The MSC/E of the force should establish a *materiel in-transit* monitoring role to ensure the accurate shipping status of selected critical assets and resources. Do not interpret this to mean involvement of the MSC/E in national responsibilities of materiel management. The MSC/E critically needs the time-sensitive ability to query national supply systems to track and forecast the arrival of decisive logistics materiel. National forces should have modern ADP systems capable of processing the quantitative analysis of all logistics functions, in addition to effectively monitoring essential items of stocks and resources. Likewise, ADP technology employed for critical functions should be compatible with other national systems as much as possible throughout the MNF. ADP and communications systems training for logistics staffs is vital for effective sustainment. The efficient transfer of logistics management information and data ensures planners have flexibility, are responsive and predictive, and meet the operational and tactical plans. Liaison teams must compensate for any technological shortfalls.

National contingents have their own support command or element to ensure their force's sustainment. This element, generically known as a national support element (NSE), must ensure that a timely and accurate logistics information management system is established between their respective nation and the MNF's MSC/E. A difficult C^2 relationship may develop between NSEs that remain under national command but are located in the multinational commander's operating area. National commanders must inform, advise, and support in all matters of combined execution. These actions allow the multinational commander to establish specific operational guidelines. The national commander must follow these guidelines within his operating area for real estate management, area security, contracting, and interface with local population and authorities, although no command relationship exists.

Legal and Financial Considerations

A commander performing a multinational/interagency mission must call upon his legal representative,[2] political-military advisors, and comptroller as urgently as he does his operations officer. Funding issues and legal issues demand much time and effort. If not fully addressed, they may either lessen mission effectiveness or cause significant bureaucratic and possible legal consequences later. Actions may include training, moving, and equipping multinational/interagency participants. Much of this falls under Title 22, Sections 505 and 506 of the *FAA*, and can only legally occur upon Presidential signature. Such actions go through DSAA, are incredibly complex, and must not be ignored if the mission is to succeed. The commander's staff judge advocate and operational judge advocate can help him understand the problems associated with multinational operations.

Congress requires details on the projected and actual costs of operations. Accurate, detailed costs are needed in dealing with multinational partners at political levels to determine how costs have been or should be apportioned. Therefore, throughout an operation, commanders must track funding, expenditure authority, accountability, costs, and support received from or provided to others. This record is necessary to determine the financial costs of the operation and to support the process of reimbursement at all levels (including international).

EXECUTION

Two general methods exist to execute multinational support: *lead nation/lead role responsibility* and *national responsibility*. The method chosen depends upon the circumstances of the alliance or coalition, the AO, and the composition of the force. The objective is to find the most efficient and politically acceptable method of providing sustainment. A key subset of these methods, *role specialization*, is also described.

Lead Nation/lead Role Responsibility

This method encompasses a nation accepting responsibility for the centralized coordination, management, and control of the contracting or procurement and provision of a broad spectrum of supplies and services for all or part of the MNF. This method can be performed by a nation (the *lead nation*) or an organization. It is normally performed at echelons above corps, although the lead nation concept can be effected within corps. The lead nation establishes a framework where other nations and organizations contribute units and resources to accomplish specified missions. In a case where a lead nation is dominant, that nation may provide almost complete CSS for the entire force. While conceptually the simplest approach, financial constraints make this option undesirable for anything short of an MRC or war.

---

[2] The US Army Judge Advocate General School produces the *Operational Law Handbook* annually. This publication addresses the wide variety of issues that commanders face while performing multinational/interagency missions. It is considered an essential reference tool for commanders as well as their legal representatives.

HISTORICAL EXAMPLE

Within UN Protective Force (UNPROFOR) in Bosnia-Herzegovina, the UK is the lead nation for vehicle recovery support to all nations along certain critical main supply routes (MSRs). While nations carried out limited self-recovery within their own zones of operations and within their own military convoys, the UK underpinned the system to ensure that vital routes remained open. UK assets also undertook responsibility for vehicle recovery outside those zones and augmented nations when possible.

If national contingents are similar in size, the nation that assumes lead nation responsibility normally requires staff augmentation and some operational support forces from participating nations. Thus, the overall CSS is more accurately described as *partially integrated.* Functions are integrated to the extent feasible depending on the specific function. Some procedures are performed solely by the lead nation, some functions are pooled and run by a multinational organization, and some may even be done by a single nation.

This approach would be most common in a coalition where some members belong to an alliance (or have some other compatibility that sets them apart from other members) and others do not. The nations with the most commonality would form the integrated structure, and other national contingents would use this structure as much as possible.

National Responsibility

Similar to the parallel mode among ABCA army members, national responsibility is the traditional approach. Each nation provides its own CSS, and the MNF headquarters monitors the support status of each unit. Any directive authority given to the MNF commander for cross-leveling supplies or services across national lines would be negotiated among contributing nations. See Figure 3-1.

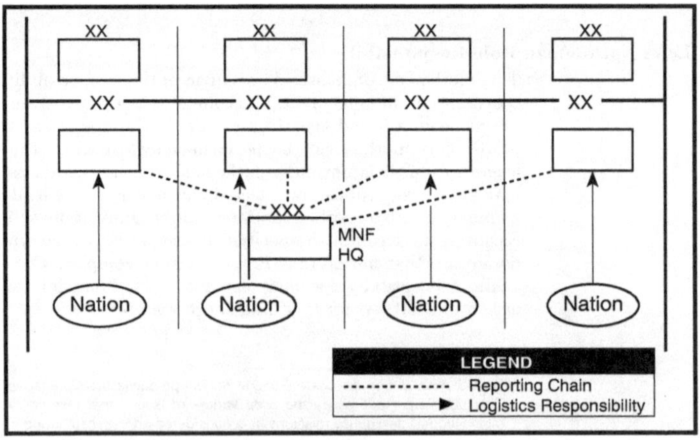

Figure 3-1. National Responsibility for Combat Service Support

Role Specialization

Role specialization is where a nation or organization assumes the sole responsibility for procuring and providing a particular class of supply or service for all or part of an MNF. Normally performed at EAC, role specialization may be executed at a lower level, depending upon the size of the MNF. Role specialization is normally used for a finite mission and time because of the great burden it places upon the nation or organization. First, unless reimbursement across national lines is agreed to before deployment of the force, the providing nation bears an unfair portion of the financial cost of the operation. Secondly, the nation or organization may use an inordinate amount of its own resources supporting others. This leads to internal supply shortfalls or delays in resupply and increases wear and tear on vehicles and personnel. Other MNF members may then have to provide other support to the partner doing the role specialization.

If properly planned and negotiated, this approach promotes greater efficiency in cases where one MNF member is already well established in the area and has contractual arrangements in place, has a unique relationship with the populace, or has a much greater capability than other nations. One example is Saudi Arabia's role of supplying fuel to coalition members during Operation Desert Shield/Storm. Another is the French role in supplying theater-level transportation and certain logistics commodities to members of UNPROFOR in the former Republic of Yugoslavia.

Optimum Method

Planners must consider a myriad of factors to determine the optimum logistical support method for a particular situation. This paragraph discusses responsibilities in categories within the respective functional areas and provides an initial framework for planners to analyze situations. The lists are neither prescriptive nor all-inclusive.

Multinational Organizations

Logistics responsibilities that can be executed by a multinational organization include—

- Movement control.
- General support of road, rail, and air transportation.
- General support of recovery operations.
- POD operations, all modes.
- Medical, roles 1 and 2, preventative medicine, and medical evacuation (MEDEVAC).
- Sanitation and refuse.
- Fresh rations and water.
- Management/operations of staging area and camp.
- Bulk petroleum, oils, and lubricants (POL) storage and distribution.
- Bulk water purification, storage, and distribution.
- Engineering resources and materials.

- Coordination of mutual assistance.
- Contracting support (of multinational functions).
- Ammunition management and security operations (possible bulk distribution).

National Contingents

Logistics responsibilities performed more effectively by national contingents include—

- Direct support of road, rail, and air transportation.
- Direct support of recovery operations.
- Maintenance operations and repair.
- Package POL.
- National ammunition procurement and distribution.
- Medical, role 3, and national preventative.
- Field rations.
- Troop support supplies and most field services.
- Major end-item replacement.
- Contracting to support national requirements. Careful coordination is required to optimize use of the available resources and minimize effects of competition in local markets.

Host Nation Support

HNS assets should, if possible, perform deployment operations, sustained operations, and closing/retrograde operations. While contingency planners cannot assume significant HNS in many theaters, they should try to find and employ existing local services and facilities upon deployment. The objectives are to preserve the logistics capability of the MNF for future operations, to strengthen and rebuild the local economy, and to positively impact the morale of the populace. HNS is the best option for—

- *Deployment operations.* Ships must be received into port and subsequently off-loaded. Local harbor pilots, navigators, and customs officials should be employed. Local storage facilities, electricity, petroleum, sanitation, and other support services should be used as much as possible.
- *Sustained operations.* MNFs work directly with local labor, training and teaching the use of new equipment and techniques. Direct results include future ideas, programs of foreign military defense sales, and commercial contracts that begin the HN's road to reconstruction. Also during this phase, MSRs should be improved by working closely with local contractors using available local resources.
- *Closing/retrograde operations.* Many additional local HNS aspects can complement the MNF: transportation assistance (buses and small transport vehicles); maintenance support operations; laundry, cleaning, bath and sanitation support; and general engineering and construction contracting operations.

Sustainment

Recent operations demonstrate future sustainment operations must be a collective responsibility with the multinational commander having sufficient authority and mechanisms over assets, resources, and forces. To require each individual nation to perform these functions separately would be inefficient and expensive and would hinder the commander's ability to influence and prioritize limited logistics resources.

MNF sustainment requires a cooperative relationship among coalition members. Whatever the method of support execution, support command representatives must meet regularly with one another and with representatives from other support sources such as contractors, near nations, and other interested parties. The representatives should assess who can (and will) provide what to whom; revise usage plans for main supply routes, airfields, and so forth; and build trust and openness among managers at a personal level. The multinational commander may choose to form a *support management cell* or similarly named organization. This cell would not be formally recognized in force structure documents. It is tailored from elements of all support organizations involved in the operation. This cell would ensure full involvement and understanding among the support managers from all of the potential sources for theater sustainment. CSS must be blended with personalities, languages, country agendas, and commitment to achieve logistics success. A formal, multinational logistics command authority will most likely not be an option and should not be expected.

Logisticians must focus on the defined military end state of the campaign plan. Other national contingents, and especially other agency participants, tend to view US support capabilities as inexhaustible; therefore, limitations on US support responsibilities are essential.

PROTECTION

Proper protection of the MNF conserves fighting potential for application at the decisive time and place. It includes actions taken to counter the enemy's firepower and maneuver by making soldiers, systems, and operational formations difficult to locate, strike, and destroy. *Operational protection* includes protecting all MNFs, bases, and LOCs from enemy operational maneuver; air, ground, and sea attack; and natural occurrences. Operational protection also includes providing operational air defense, safeguarding operational forces in campaigns and major operations, and employing operations security (OPSEC)—to include providing physical and personnel security, and conducting deception.

AIR DEFENSE

Operational protection includes protecting joint and multinational air, space, land, sea, and SOF; bases; and LOCs from enemy air and missile attacks. Dedicated, ground-based, all-weather, 24-hour ADA units execute the bulk of the theater missile defense (TMD) and AD force protection missions. Army ADA protects forces and selected geopolitical assets from aerial attack, missile attack, and surveillance during war and throughout the full range of military operations, to include the transition from war to peace. After transition, ADA units typically remain in overseas theaters for extended periods. Significant considerations for the employment of ADA in theater operations include its role in joint and multinational operations.

The Army Air and Missile Defense Command (AAMDC) is the theater-level organization conducting TMD/AD operations for the JFC and controls Army AD assets. It is linked to joint and multinational organizations and coordinates and integrates TMD/AD operations with the airspace control authority (ACA), area air defense commander (AADC), and JFACC as required. Short-range air defense (SHORAD) elements are integrated throughout the theater of operations and provide force protection during all phases of force-projection operations.

The contributions of all services to theater missile and AD offensive and defensive tactics ensure engagement of all applicable targets. These offensive and defensive tactics cover all aspects of active and passive defense measures throughout the theater. During passive defense operations, AD elements conduct counterreconnaissance, surveillance, and target acquisition/ unmanned aerial vehicle operations to enhance early warning.

Counterair Operations

Counterair operations are conducted to attain and maintain a desired degree of air superiority by destroying or neutralizing enemy forces. Joint and multinational counterair operations include both offensive and defensive measures taken against enemy air threats. Offensive counterair (OCA) operations destroy, disrupt, or limit enemy air threats as close to their source as possible. Defensive counterair (DCA) operations are conducted primarily to counteract enemy air offensive actions to nullify or reduce the effectiveness of hostile air attacks.

AD forces conduct DCA operations using both active and passive measures. Active DCA operations include ADA, EW, chemical (smoke), and air elements to disrupt or destroy airborne enemy aircraft, missiles, and other aerial vehicles that pose attack and surveillance threats. Passive DCA measures such as cover, concealment, signature reduction, smoke operations, and deception frustrate enemy targeting efforts and minimize the effects of enemy attacks.

Theater Missile Defense Operations

TMD operations may be required within the context of an alliance, coalition, or other international agreement. Commanders must consider those areas peculiar to multinational operations, such as force capabilities and disparities, information and equipment security levels, and procedural and organizational differences as they influence the ability to achieve unity of effort. Commanders must identify differences in doctrine, training, equipment, and organization and consider such differences when determining multinational interoperability requirements for employing forces. Leaders must approve command relationships among the elements of the alliance or coalition. The objectives of TMD are—

- To reduce the probability of and to minimize the effects of damage caused by theater ballistic missile attack.
- To detect and target theater missile platforms.

- To detect, warn of, and report theater missile launch.
- To coordinate a multifaceted response to theater missile attack.
- To integrate TMD with other combat operations.

TMD has four operational elements—active defense, passive defense, attack operations, and command, control, communications, computers, and intelligence (C^4I). The Army contributes to all four. Active defense engages missiles and enemy aircraft armed with air-to-surface missiles in flight. Passive measures reduce the vulnerability of critical forces and assets to theater missile attack. Attack operations are conducted to prevent the launch of theater missiles. C^4I is required to coordinate and integrate the defense against the theater missile capability with other services and nations. FM 100-12 and FM 44-100 provide detailed information on TMD and AD operations.

Nuclear, Biological, and Chemical Defense

A detailed warning and reporting system, supported by a viable communications network and a standardized theater alarm system, must be established. This system will inform all MNF units of impending or actual employment of NBC weapons. The MNF should establish standards for levels of protective posture, detection, and decontamination. All MNF units should understand these criteria to ensure overall force protection. Protective and decontamination assets should be shared, and NBC reconnaissance assets centralized or at least coordinated to cover all forces and important areas. A set protocol for handling NBC samples must be developed because of the political implications of confirming the use of an NBC weapon. The large number of casualties that results from the use of NBC weapons will tax the medical community, but tremendous opportunities for standardization and interoperability exist for their treatment.

Deception

Care must be taken to ensure that our deception of an opponent does not affect MNF and interagency partners. To prevent our deception effort from confusing the MNF, FAOs, LNOs, SOF, and State Department personnel should be used to ensure the correct messages are sent. Language cannot be directly translated because of nuances and phrasing.

Chapter 4

Tactical-Level Considerations

...difficulties are likely to arise unless higher commanders and staffs possess a knowledge, not only of the organization and staff methods of their allies, but also of the organization and general principles of the tactical employment of allied subordinate formations.

Notes from Theatres of War, London War Office
October 1943

The principal focus of this manual is on multinational army operations at the operational level. However, it is worthwhile to examine the impact of multinational operations on tactical-level decision making as well. In the vertical linkage between the operational and tactical levels, the lines of distinction between levels are sometimes unclear. Commanders often find that while operating at their level they are simultaneously required to understand the higher intent and to plan two levels down. Therefore, commanders must understand tactical-level multinational matters. Tactical cooperation requires more precision since it deals with immediate combat actions. Among the disparities that adjacent and supporting commanders must reconcile are differences in tactical methods and operating procedures; differences in using other service capabilities, such as CA, varying organizations, and capabilities of units; and differences in equipment.

As they are at the operational level, tactical-level considerations are grouped into different functions. They include maneuver, intelligence, firepower, AD, CSS, C^2, and mobility and survivability. The intent here is to examine them from a multinational perspective and consider how each might be uniquely affected when applied in a multinational tactical environment.

MANEUVER

Tactical maneuver is characterized by the employment of forces on the battlefield through movement and direct fires in combination with fire support, or fire potential. This achieves a position of advantage over enemy ground forces to accomplish the mission. Tactical maneuver includes direct fire systems, such as small arms, tank guns, and attack helicopter fires, but it does not include indirect fires that are included under fire support.

The multinational commander must look at differing capabilities among partners when developing the maneuver portion of his plan. When operating under a lead nation concept with armies arrayed as national entities, imposing the lead nation's scheme for tactical maneuver on all participants would not suffice. Differences in TTP and disparity in equipment capabilities render such an approach unworkable. The use of liaison teams and an experienced staff to determine capabilities and differences of each partner cannot be overemphasized.

The multinational commander must familiarize himself with the capabilities and procedures of the armies under his command. This allows him to develop a maneuver plan that capitalizes on the strengths of each partner. Familiarity with the capabilities of all partners also enables the multinational commander to reach a sound decision as to the proper task organization. For instance, only one nation's army may have a strong attack helicopter capability, and this capability is the key to success. The multinational commander might establish command and support relationships that provide an effective attack helicopter capability throughout the MNF.

INTELLIGENCE

Tactical intelligence is required to plan and conduct tactical operations. Intelligence-preparation-of-the-battlespace (IPB) is a continuous process that assesses the situation by evaluating the threat and describing the environment and its effects on the force. IPB provides a comprehensive projection of the situation for the decision maker and the synchronization of that information with the tactical commander's intent and plan. It is derived from an analysis of information of the enemy's capabilities, intentions, vulnerabilities, and the environment.

There is no single intelligence doctrine for multinational operations. Each coalition or alliance must develop its own procedures. Wherever possible, the multinational command and national forces' intelligence requirements, production, and use should be agreed upon, planned, exercised, tested, and replanned prior to operations. The best way for the J2 and his staff to prepare in this area is to anticipate what information and intelligence must be and may be exchanged and obtain required authorizations to enable these exchanges. Architecture and procedures appropriate to the equipment and liaison elements of the participating countries are then established. Collecting, disseminating, and sharing intelligence in the multinational tactical environment is difficult because each nation imposes its own operational and electronic protection measures on their forces. Often, these security measures preclude open sharing of information between partners and force the multinational commander to adopt measures ensuring that critical intelligence is accessible to all participants. This may involve requesting special dispensation from certain security restrictions. The senior US commander may modify any restrictions or procedures if they appear to jeopardize the mission or friendly forces. When intelligence cannot be shared among coalition nations, it should be provided after it is sanitized by effectively separating the information from the sources and collection methods.

In the multinational arena, indigenous populations provide a valuable source of tactical intelligence, particularly regarding terrain, weather, and information about local political relationships. This resource must be fully exploited where possible. The multinational commander must also tap the experience of multinational partners who have previous experience in the operating area. This is especially important if a nation's forces have a good working relationship with the local populace. Use of previously trained US personnel can also facilitate the use of HUMINT resources.

Regarding intelligence-gathering and dissemination systems, incompatibility of equipment is likely to be a problem for the multinational commander. If coalition and allied processing and communications systems are not compatible with US systems, commanders may choose to cross-level resources or establish special command and support relationships, such as use of intelligence support elements (ISEs). Liaison is often the key to obtaining vital information quickly.

Besides equipment incompatibility, internal staff structures and methods are often incompatible as well. For example, US staff cells typically surpass their counterparts in number, rank, and radio nets. Other national contingents sometimes perceive that US intelligence sections disseminate a great deal of unprocessed information, not intelligence.

A significant area of difference is in the US reliance on technical intelligence-gathering means to the detriment of other means. Other nations focus on and have more ground reconnaissance forces than a US force. This is an excellent way to cover gaps in US capabilities, such as the lack of scouts at the brigade level. Intelligence and operations officers must carefully research and employ all available assets across the MNF.

FIREPOWER

Tactical firepower has two major functions:

- The collective and coordinated use of target acquisition data, indirect fire weapons, armed aircraft (less attack helicopters), and other lethal and nonlethal means against ground targets to support maneuver force operation.
- All measures used to nullify or reduce the effectiveness of attack by hostile aircraft or missiles after they are airborne.

Tactical fire support includes artillery, mortar, and other nonline-of-sight army fires, naval gunfire, CAS, and EW while AD includes all weapons systems with potential to engage aerial targets.

FIRE SUPPORT

Synchronization is key to effective tactical fire support in a multinational environment. The MFC must organize his staff to ensure the fire support element can synchronize the fire support capabilities of all friendly forces. This serves a twofold purpose. First, it provides the capability of applying the most effective fire support to a given target in the most timely fashion. Second, bringing multinational fire support under centralized control as much as possible provides another safeguard against fratricide.

The multinational commander must ensure that the fire support annex to his OPLAN provides detailed information regarding the means by which the multinational fire support element operates. As the focal point for fire support, this element must have the means of controlling all fire support at its disposal. Usually, this requires representation from all member nations. It would be the responsibility of these representatives to coordinate fire support matters between the multinational headquarters and their respective army fire support activities.

If time permits, the MFC should direct the establishment of an SOP for fire support. Simplicity should be the foundation of this document. If establishment of a formal SOP is not practical, the fire support annex to the OPLAN must be comprehensive enough to provide detailed information as to how this aspect of the operation comes together.

Air Defense

Multinational AD assets are integrated and managed at the operational level to achieve a balanced mix of AD forces available to the operational-level commander. The operational-level commander establishes both positive and procedural measures to minimize mutual engagement interference and establishes readily identifiable means to protect friendly aircraft. Additionally, the operational-level commander provides the tactical-level commanders guidance on priorities of effort to provide maximum protection and use of the tactical commander's AD assets.

The tactical-level commander must focus his AD assets on protecting his formation, in line with the mission. There are no specific AD concepts for multinational operations, yet there are actions that must be emphasized between tactical-level multinational AD forces.

Perhaps the single most important action is the additional coordination required between allied AD forces in order to better understand one another's capabilities and procedures. This may include—

- Establishing communications nodes.
- Sharing tactical and intelligence information.
- Developing/confirming early warning procedures for respective sectors.
- Establishing liaison teams for the coalition's AD units.
- Coordinating other actions that promote a better understanding of the multinational commander's intent.

Development of cross-service capabilities is necessary.

Airspace C^2 considerations are critical aspects of the AD battlefield operating system. Commercial aircraft, helicopters, and air force, navy, and marine corps aircraft use the same airspace.

Air and missile defense at the tactical level entails all measures designed to nullify or reduce the effectiveness of attack by hostile aircraft or missiles. It includes all weapons systems with the potential to engage aerial targets.

The tactical-level commander focuses on the lethal engagement of air and missile targets, conducts nonlethal engagement of air and missile targets, and denies the enemy use of airspace. The tactical commander has ADA, joint, and multinational active defense systems to accomplish these tasks. Tactical-level commanders ensure that AD operations are fully integrated into the land component commander's plan of operations for a coordinated effort of providing force and asset protection. The prioritization of assets for protection is critical to successful accomplishment of AD operations.

During multinational AD operations, many considerations must be addressed. Among them are weapons systems types and capabilities, logistics requirements (arming, fueling, fixing, moving, and supporting soldiers), interoperability connectivity, communications, and leadership. All levels of command are responsible for ensuring that TMD/AD operations are efficiently and effectively implemented. Specific consideration must be given to integrating similar or like systems having distinctly different systems capabilities. To identify shortcomings and prevent deficiencies, training and rehearsals must be integrated into daily operations, since variations in systems can hinder effective operations. To better understand each nation's capabilities and procedures, the establishment of communication nodes, the sharing of tactical and intelligence information, the development of alert and early warning systems, and the provision of liaison teams (linguists) promote unity of effort.

Airspace C^2 is critical to successful air and missile defense operations since both commercial and military aircraft may operate within the protected airspace. FMs 100-103, -1, and -2 provide detailed information on Army airspace C^2.

HISTORICAL EXAMPLE

On 12 January 1991, in response to a growing tactical ballistic missile (TBM) threat in the Persian Gulf, the Secretary of Defense authorized the deployment of two USAREUR Patriot batteries to Turkey to provide TBM defense for Incirlik Air Base. By 22 January, six of the eight launchers were in place and operational with 43 missiles on hand. The United States and Israeli political authorities agreed to deploy Patriot units to counter TBM threats (in the form of Soviet-built SCUD missiles) to Israel. Shortly after the war began, Iraq attacked Tel Aviv and Haifa, Israel, with an extended range variant of the SCUD B missile. A direct Israeli military response to these attacks might have weakened the commitment of coalition Arab members to Operation Desert Storm. Task Force Patriot Defender, created from the 32d Air Defense Command (USAREUR), deployed to Israel to provide antitactical ballistic missile defense of priority Israeli assets and to provide training and maintenance support for the two newly formed Task Force Patriot batteries. Patriot units from the 32d Air Defense Command were ordered to deploy. The task force was operational, and ready to fire. A second deployment of two more batteries to Israel began on 23 January and was completed and operational by 26 January.

DOD Final Report to Congress

Conduct of the Persian Gulf War, April 1992

COMBAT SERVICE SUPPORT

Tactical CSS sustains the tactical commander's ability to fight battles and engagements by providing the right support at the right time and place to units in the combat zone. The focus at the tactical level is on manning and arming tactical units, fixing and fueling their equipment, and moving and sustaining soldiers and their systems. The multinational commander must thoroughly integrate CSS with the concept of operations during planning.

The levels of CSS compatibility in multinational army CSS have a great influence on the functions and actions of commanders. As a result, planning staffs must evaluate the level of compatibility and commonality between participating nations and, where situations permit, agree on—

- The nations responsible for providing support functions for the MNF.
- The task organization of the CSS units to support the MNF.
- The procedures and methods to provide the support.

Bilateral and multilateral agreements must clearly define command and support relationships, national responsibilities, and multinational responsibilities. The multinational CSS coordinator must be aware of all agreements between and among participating nations to integrate support and adapt it to the combatant commander's intent.

Multinational compatibility, commonality, and cooperation must be reflected throughout the multinational CSS command with the objective of blending manpower, equipment, and resources into the most effective combat force possible. However, many multinational sustainment concerns will be encountered across the full range of CSS—from providing soldiers with food, water, equipment, and clothing (*resources* and *assets*); to providing health care, ammunition, fuel, transportation, and movement operations; and finally, to performing maintenance, recovery, and repair parts management (*functional support areas*).

At the MNF headquarters, the support focus should be on—

- Measuring the requirements for executing the campaign plan.
- Providing advance estimates of these requirements to national units.
- Ensuring that proper controls are in place to deconflict and permit movement and processing of combat power to units.

Actual execution of tactical support to MNF members should be decentralized.

An important aspect of operational sustainment is planning for future operations. Commanders must be aware of the current and projected requirements of their own force and projected logistical requirements of various members of the MNF they may have to support.

At the respective nations' tactical level, multinational CSS C^2 would be executed through and by a national CSS unit. The CSS unit should be a flexible organization able to deploy units and materiel throughout the operating area. If the scale of the operation or the length of the LOC dictates, the CSS unit should be able to split into two parts—a rear support unit (RSU) and a forward support unit (FSU). The RSU would execute control of national resources and assets in the rear area and in proximity of the POD. In the rear, depending on the situation, the RSU may collocate with the CSS unit. Also, the FSU would establish a position close to the operating area with more mobile units. In addition, the FSU would provide necessary logistics support to forward logistics units and forces. The FSU may receive selective logistics support from a lead nation tasked to provide specific logistics support in the operating area of the FSU.

HISTORICAL EXAMPLE

On 17 January 1991, the 6th Light Armor Division (French) was assigned under TACON of the XVIII Airborne Corps. It replaced the heavier 1st US Cavalry Division within the XVIII Airborne Corps and provided a faster, less logistically dependent force with which the corps could execute its offensive mission.

The French arrived on the west coast of Saudi Arabia at Yanbu and had been operating independently until 17 January with a national LOC back through that test coast city to France.The XVIII Airborne Corps LOC ran east to Ad Dammam. With different in-place and effective supply systems, planners decided to collocate logistical units but not to further integrate the supply process. The US provided essentially only water and fuel for French forces during the operation. Class I (water) support was provided through a supply point system until G-Day.The corps brought water forward to the logistical base in the tactical assembly areas. French vehicles picked up the water and carried it forward to their division support area. Once the attack began, corps provided water support directly to the division support area for the French, thus releasing the French's organic transport for internal distribution.

Class III bulk was also a supply point system prior to G-Day and continued as such throughout the conflict. The French division had a fuel system supply point with collapsible bags at their division support area. This allowed them to build up a supply of fuel for several days of operations. Fuel tankers from corps and higher provided fuel to the corps fuel supply point, and division assets carried it forward to their units.

CIVIL AFFAIRS

CAs are the activities of a commander that establish, maintain, influence, or exploit relations between military forces and civil authorities, both governmental and nongovernmental, and the civilian population in a friendly, neutral, or hostile AO to facilitate military operations and consolidate operational objectives. CA may include performance by military forces of activities and functions normally the responsibility of the local government. These activities may occur before, during, or after other military actions. They may also occur, if directed, in the absence of other military operations.

Prior to hostilities and in peacetime, CA units may help a nation create or improve its own infrastructure, thereby precluding grievances from flaring into war. During hostilities, the primary role of CA is to ensure that civilians do not interfere with operations. The secondary role is to protect and care for civilians in a combat zone and assist the commander in fulfilling his legal and moral obligations to the civilian population.

TRANSPORTATION

By synchronizing the employment of water, rail, road, and air transportation assets available to the MNF, the effectiveness and efficiency of moving supplies, equipment, personnel, and EPW are enhanced. In most multinational operations, a centralized authority should be established. This authority would allocate space, set the order of deployment, and deconflict requirements to support the national elements of the multinational formations. Several methods can be used to organize and manage the MNF's transportation requirements. They include—

- Establishing a multinational movement control agency or similar organization.
- Consolidating general support transportation assets under the control of one nation.
- Operating sea/air terminals by a single nation, if appropriate.

SUPPLY

Significant economy and savings of the MNF's POL resources and efforts can be achieved by consolidating bulk fuels and refining storage and distribution requirements. The result is an increase in tactical flexibility and maximum use of fuel transport equipment and storage capabilities. Planners should consider forming a multinational petroleum office, similar to the joint petroleum office, as soon as possible. To reduce redundancy, nations should agree to have one nation operate storage facilities and pool transport assets under another nation's control. Planners must ensure that couplings are interoperable between national assets.

From small arms ammunition through artillery shells and charges, from different storage and transportation requirements, to national policies and procedures for redistribution, the ammunition requirements of an MNF are tremendous. Planners can assume little in this area. For example, different nations use different powder for their 155mm artillery shells, which produces different firing tables. Various national storage, transportation, surveillance, materials handling, and issue and receipt policies and procedures require significant investigation and revision for effective use during a multinational operation. Planners must develop a matrix of weapons systems that identifies compatibility of weapons and interchangeability of munitions to support the force. NATO's STANAG 2928[1] provides a baseline to accomplish this task. Where feasible, consideration should be given to establishing collocated ammunition storage facilities. The above actions reduce redundancy and economize support assets.

PERSONNEL SERVICES

Due to the nature of PSS, multinational PSS assets should be collocated. However, individual national application of PSS remains a national responsibility. This allows for ease of coordination between the national elements of the MNF even though personnel functions are not amenable to consolidation.

Finance operations and resource management focus on supporting local procurement efforts and on funding and tracking operations costs. Finance elements help access HN (private sector) and other support by paying for contracts and by providing cash to paying agents to make local purchases for immediate needs of the force. Finance personnel also provide military pay support and provide limited support to other services, such as funds for

[1] *Allied Operational Publication 6 (G), Land Forces Ammunition Interchangeability Catalogue in Wartime*, updated annually.

paying agents and cashing checks as agreed upon and cashing checks and exchanging currency for individuals. Resource managers focus on obtaining obligation and expenditure authority and on tracking the costs of the operation. Finance elements provide essential input into the accounting systems to support cost capturing. Accurate, detailed costs are needed for reporting to Congress and for government dealings with multinational partners to determine how costs have been or should be apportioned.

Field Services

Mortuary affairs, airdrop, water production and distribution, and laundry and shower services lend themselves to central planning and distribution. Rations have the potential for some degree of mutual support as long as nutritional and caloric requirements are met and significant national taste and/or religious concerns do not exist. If taste and/or religion are major factors, one nation could provide the basic ration with affected nations providing supplements to meet national requirements.

Combat Health Support

While most MNFs deploy with some form of combat health support, wide disparities always exist in the quality and quantity of medical services provided by each of the participants. The consolidation of multinational medical assets should be evaluated to maximize availability and effectiveness of combat health support. Coordination between multinational medical units is essential for effective MEDEVAC support, specialized medical services, and hospitalization of force members. The US, as with most other nations, retains control over medical care to its forces, except in life-threatening situations. When an MNF member receives emergency treatment from another nation, the patient is first stabilized. Then, he is transferred via MEDEVAC to a national medical treatment facility in or near the operating area. Depending on the mission, combat health support requirements may differ. Combat health resource requirements may include trauma treatment for casualties, preventive medicine for displaced civilians, or veterinary support for local farmers

HISTORICAL EXAMPLE

Collocated medical facilities allowed US MEDEVAC helicopters to provide lift to both US and French evacuees to a single point for both hospitals. A centralized triage point provided for segregation of injuries as evacuees were removed from the helicopters with both US and French medical personnel on site.

With both US VII Corps and XVIII Airborne Corps moving to tactical assembly areas simultaneously, US and Saudi heavy equipment transporter (HET) assets were overtaxed. The French offered the use of 42 of their HETs. Commanded by a French officer, the French transports reported to the XVIII Airborne Corps Transportation Control Center (TCC) at An Nuayriyah for direction. They then transported the US armor to the corps tactical assembly areas in the vicinity of Rahfa and returned to the TCC for their next mission. This support continued until the move of the corps was complete.

MAINTENANCE OPERATIONS

The greatest challenge for logistics planners, without benefit of common equipment, is the capability to achieve mutual support in the area of maintenance operations. Again, a matrix of equipment must be developed to determine where mutual support can be provided. Logistics planners should also consider collocating maintenance organizations, sharing common test equipment, and consolidating or integrating recovery and evacuation assets.

HOST NATION SUPPORT

Planners should consider centralizing HNS expertise (legal, financial, procurement, contracting, and administrative). This would ensure the force's total requirements are known, prevent competition among partners, and allocate support based on command priorities that best support the operational objectives. CA units have this type of expertise available. The MNF's total logistics requirements are known to prevent competition among partners. The allocation of this support is based on command priorities that best support the operational objectives.

In the area of HNS, the multinational commander must know the amount and type of HN or counterpart logistical support available to him in his AO. While some operating areas have well-established HN agreements, the multinational commander must plan and prepare to coordinate required support in theaters where established agreements are not present.

Examples of support that can be provided by the HN include—

- Supply (to a limited degree).
- Maintenance (to a limited degree).
- Recovery and evacuation.
- Transportation.
- Hospitalization and MEDEVAC.
- Construction.

The multinational commander should plan and pursue any form of this support that the HN can provide before or immediately upon arrival in his operating area.

COMMAND AND CONTROL

Tactical C^2 in a multinational environment is the exercise of authority and direction by a properly designated multinational commander over assigned forces to accomplish the mission. Multinational C^2 activities are performed through an arrangement of personnel, equipment, facilities, and procedures to plan, direct, coordinate, and control forces and operations.

The most obvious C^2 considerations in a tactical multinational environment are differences among partners in TTP and equipment with procedural differences causing the most friction. The multinational commander must recognize, however, that the key purpose of C^2 remains unchanged. That purpose is the process of generating and applying combat power decisively.

US C^2 procedures often differ greatly from those employed by other nations. Many of our multinational partners categorize our C^2 system as decentralized (i.e., orders emanate from the bottom), while they feel that their own C^2 procedures are centralized (i.e., orders emanate from the top). Additionally, they see US procedures being too lengthy and indirect, while they view their own systems as shorter and more direct. The US use of multiple fragmentary orders (FRAGOs), from the basic operations order (OPORD), is a potential source of confusion that US staffs must consider. Also, the size of US headquarters elements tends to overshadow their counterparts and leads to confusion over where they can best tie into the US C^2 apparatus. Some common techniques, such as walk-through rehearsals,[2] are extremely effective in ensuring common understanding among multinational partners. They are the easiest medium to ensure that all players understand the commander's intent and are properly synchronized in all phases of an operation. Walk-through rehearsals provide an excellent opportunity for the various national commanders to ensure that national representatives fully understand capabilities of other nations. For example, discussion of a defensive plan would bring out different intelligence sources, such as sound ranging, differing ground reconnaissance assets, and artillery ranging, that planners may have overlooked.

Another C^2 consideration not fully appreciated or understood is selection of a C^2 language. English is not automatically the C^2 language. Mistakes or misunderstandings happen during the translation of the C^2 language into an ally's native tongue. Language problems are solvable—but not quickly, and the process requires tremendous dedication of resources.

As in unilateral operations, METT-T impact upon the C^2 architecture of an MNF. In the multinational environment, however, the variances in TTP and equipment provide additional considerations that must be factored in. If armies have little or no previous experience operating with one another, the multinational commander must devise a command and staff arrangement that capitalizes on the strengths of multinational armies. The actual command structure may be imposed on the commander by higher military or political authority. However, the commander must retain the flexibility to adjust the architecture to ensure harmony and unity of effort. If control mechanisms are so dissimilar as to preclude interoperability, the commander must create an operations center comprised of personnel and equipment that enables him to control all forces from at least one central location. One implied requirement here is requisite linguistic ability for personnel assigned to the operations staff. Another *given* is that liaison teams and personnel must be employed by all partners to augment the C^2 effort. Planners must prepare for rapid changes in mission that alter the type and priority of support needed. Interoperability of communications and a standardized report system are critical to the success of the operation.

As for command relationships and human factors (political, cultural, and economic issues and the impact of personalities on multinational operations), the same considerations discussed at the operational level apply at the

[2] *Sticker drills* have proven very effective in communicating the commander's intent within an MNF.

tactical level. Early agreement upon C^2 issues facilitates agreement upon other tactical-level functions.

MOBILITY AND SURVIVABILITY

Tactical mobility and survivability provide a force freedom of movement relative to the enemy, while allowing the force the ability to fulfill its primary mission. Tactical mobility and survivability include those protection measures the force takes to remain viable and functional, despite the effects of enemy weapon systems and natural occurrences. Consideration of functions such as deception, OPSEC, security, and NBC activities can help avoid enemy detection and reduce the effects of enemy weapons.

Commanders must recognize the distinction between the function of *maneuver* and the function of *mobility and survivability*. Specifically, maneuver functions pertain to movement for positional advantages. Mobility and survivability functions pertain to enhancing friendly movement or protecting friendly forces from the effects of enemy weapon systems and natural occurrences.

Of significance to the multinational commander is his knowledge of the capabilities (strengths and weaknesses), in terms of mobility and survivability, of the MNFs under his control. Again, differences in TTP become principal considerations with regard to tactical mobility and survivability. The assessment and inclusion of mobility and survivability functions in tactical plans are standard practice for most nations in unilateral operations; yet, when considered from a multinational perspective, these same functions require commonality of procedures to be effectively planned and executed.

MOBILITY FUNCTIONS

TTP and tactical plans must provide freedom of movement for MNF on the battlefield without delays caused by terrain or obstacles (to include dislocated civilians). Mobility functions that require special multinational considerations include—

- Facilitating movement.
- Reducing obstacles.
- Employing tactical bridges.
- Upgrading LOC capabilities.
- Removing and disposing of dislocated civilians.

These functions are tailored to enable an MNF to maintain its mobility and ability to fulfill its primary mission while delaying, channeling, or stopping offensive movement by the enemy. An example of these functions is found in engineer support of an MNF, which impacts significantly on MNF mobility. Engineer task-organization decisions require an assessment of both US engineer capabilities and other nations' engineer capabilities. For example, an allied force may have a better breaching capability but lack survivability equipment. The multinational commander may task-organize to use these strengths. Commanders need to address interoperability, logistics, language, and doctrinal concerns. Early in the planning phase, commanders should establish procedures to address these issues. Also, commanders should determine the C^2 system that most effectively enhances the mobility, countermobility, and survivability skills of engineer units. Reporting procedures, national interest, intent, and assessment of unique engineering tasks or problem areas need to be clarified among allies.

The multinational commander may clarify problem areas by directing an in-depth engineer estimate be performed on the mission and capabilities of the MNF. The estimate should determine—

- The engineer capabilities each ally provides.
- The engineer requirements of each force in the MNF.
- The available HN assets and capabilities.
- The task-organization that makes the best use of the strengths of each allied engineer force.

SURVIVABILITY FUNCTIONS

The ultimate objective of all commanders is to accomplish the mission with minimal loss of personnel, equipment, and supplies. Commonality of key survivability functions is of particular concern to the multinational commander simply because failure to address and establish common procedures for MNF could result in mission failure. Survivability functions affect every aspect of tactical plans. Key survivability functions that the multinational commander must consider are—

- Battlefield hazard protection used to protect friendly forces on the battlefield by reducing or avoiding the effects of enemy weapons through the use of armor, NBC protection, fighting positions, overhead cover, warning devices, and so forth.
- Actions such as identification, friend or foe (IFF) to prevent fratricide and friendly fire damage to MNFs and other marking or procedures to positively identify friendly forces.
- OPSEC procedures such as signal, transmission, information and communications security, concealment techniques, camouflage, noise and light discipline, counterreconnaissance, and physical security controls are used to deny the enemy information about friendly capabilities and intentions by identifying, controlling, and protecting indicators associated with planning and conducting multinational operations.
- Deception in support of multinational tactical operations that mask the real objectives and delay effective enemy reaction by misleading the enemy about friendly intentions, capabilities, objectives, and the locations of vulnerable units and facilities. An effective deception program employs both electromagnetic and physical means. The multinational commander should be prepared to employ physical deception tactics such as demonstrations, feints, ruses, displays, and deception smoke screens. All of these are intended to prevent the enemy from learning the intentions of friendly forces.

HISTORICAL EXAMPLE

XVIII Airborne Corps began addressing fratricide reduction measures immediately upon assuming tactical control of the 6th Light Armor Division (French) during Operation Desert Shield. All Arab coalition forces screening to the north of the corps defense positions had a mix of Soviet, European, and US equipment, much of which was identical to Iraq's.

In addition to identification panels, inverted "V," and light markings addressed in the SOP, the corps and the 6th French recommended the inclusion of three white "invasion" stripes on French aircraft to clearly identify them as friendly. Also, the 6th French established a schedule of visiting US combat units with their ground and air systems to help visually familiarize US gunners and vehicle commanders with French equipment. The success of these efforts was confirmed by the 101st Air Assault Division's officers and NCOs who had French aircraft or vehicles on their flanks but did not engage the French due to the system's markings and soldier familiarity with French systems.

To reduce the possibility of fratricide, the hard surface road (MSR TEXAS) was used as the boundary between US and French ground units. Additionally, the 2d Brigade, 82d Airborne Division, and its adjacent French units, the 3d Rima and the 4th Dragoons, conducted detailed coordination on their respective maneuver plans following the map exercises. They also exchanged bilingual liaison teams at the battalion level to accelerate cross-battalion coordination in combat. Also, visual contact was maintained between adjacent French and US units throughout the operation, often with French and US vehicles or soldiers only 50 meters apart.

Chapter 5

Planning and Preparation

You must be prepared...to accept minor inefficiencies as long as that is promoting the great and common purpose...You should not try to change ideas and concepts on the part of some subordinate of a different nationality because you disagree with him. If you can achieve the great overall unity of purpose that inspires loyalty, inspires teamwork, never bother your head about things in seeking perfection because too many difficulties can arise out of minor irritations and frustrations. You must not lose your sense of humor because if you do your allied command will blow apart.

General Dwight D. Eisenhower, in a speech on "Problems of Combined Command"

Up to now, this manual has focused on what needs to be done prior to conduct of a multinational operation. This chapter contains the two key steps necessary to synergistically meld the previous chapters' requirements. Planning and preparation are the two major facets for successful execution of multinational operations. The glue that binds a multinational operation together is the commander's ability to understand and mesh each counterparts' capabilities, personal and professional habits, training backgrounds, relevant national characteristics, and national goals into unity of effort. This unity requires the commander to clearly articulate his intent, so that each nation forms the same picture of the end state and the rules governing engagements. Attempts taken to unify the effort between leaders contribute immeasurably to the success of the alliance or coalition.

SETTING THE STAGE

Prior to any operation, the playing field must be established. In multinational operations, defining the limits of an operation quickly determines who joins the coalition, who supports the coalition, and who opposes the coalition. Therefore, for a commander to become an MNF commander, he must clearly articulate the strategic *end state*, overall *intent*, and basic *ROE* to each of the participating nations.

DETERMINING AN END STATE

Success is more likely if nations agree on a strategic end state with well-defined termination and exit conditions to achieve the military end state. These conditions help prompt the decision to end an operation, and all participants in the MNF should agree to these conditions. Exit conditions are critical to the transfer of responsibility from the MNF to another authority—national, UN, other regional political bodies, or overall termination of the operation. As such, the UN Security Council resolutions may impose these conditions. The UN or political leadership sets the strategic end-state conditions before commitment of forces.

Understanding the Commander's Intent

The commander's intent is a concise expression of the purpose of the operation and describes the desired end state. Subordinate and supporting commanders must understand the *higher* commander's intent. The intent statement helps subordinates pursue the desired military end state without further orders, even when operations do not unfold as planned.

Given the inherent language difficulties found in many multinational environments, the multinational commander's intent must be clearly and simply stated when issuing orders to the MNF. When US forces find themselves in a multinational command structure, US military commanders must make every effort to ensure that the senior multinational commanders recognize the importance of the commander's intent and a clearly defined end state.

Understanding the Rules of Engagement

ROEs are a critical factor that must receive close attention during the formation of the MNF. Each nation comes into the MNF with a different national ROE reflecting that nation's reason for entering the coalition. Some national ROEs will be relatively free of constraint while others will have very limited authority to act or react to a situation. The MNF commander cannot be expected to command multiple forces with different ROEs successfully. These differences must be harmonized, and one standard with common definitions established for the entire MNF. For the individual soldier to understand and implement ROEs, they must be clear and simple.

PLANNING

Planning for US participation in multinational operations must start well ahead of the actual operation and may use generic plans as a framework to build the specific plan around. To ensure this effort incorporates all major facets of an operation, the following phases must be adequately addressed in the planning stages:

- Predeployment.
- Deployment.
- Leader preparation.
- Sustainment.
- Transition operations.

National Security-level and National Military-level Planning

Military advice to the NCA in the early stages of determining the strategic end state, objectives, and composition of multinational operations is critical to a successful operation. The senior Army commander and his staff should take every opportunity to ensure political leaders are fully aware of the capabilities and limitations associated with Army forces used in a multinational operation and the time required to successfully plan and prepare for achieving the military end state. Multinational operations must be carefully thought through so that pressures of execution do not fracture the alliance or coalition.

Informing the State Department of which countries US forces are most familiar and compatible with aids them in coalition recruitment. It is also an excellent example of how military planners can assist political leaders. See Figure 5-1.

CAMPAIGN PLANNING

Thorough campaign planning ensures unity of effort across different nationalities, NGOs, and PVOs. US campaign planning processes are simple enough for other nations to agree to and understand. Similarly, other nations may contribute process and content to improve planning. Habitual relationships in peacetime or sufficient training time before operations begin allows US military planners time to familiarize others with the key points of the process and builds consensus on the approach to the particular operation.

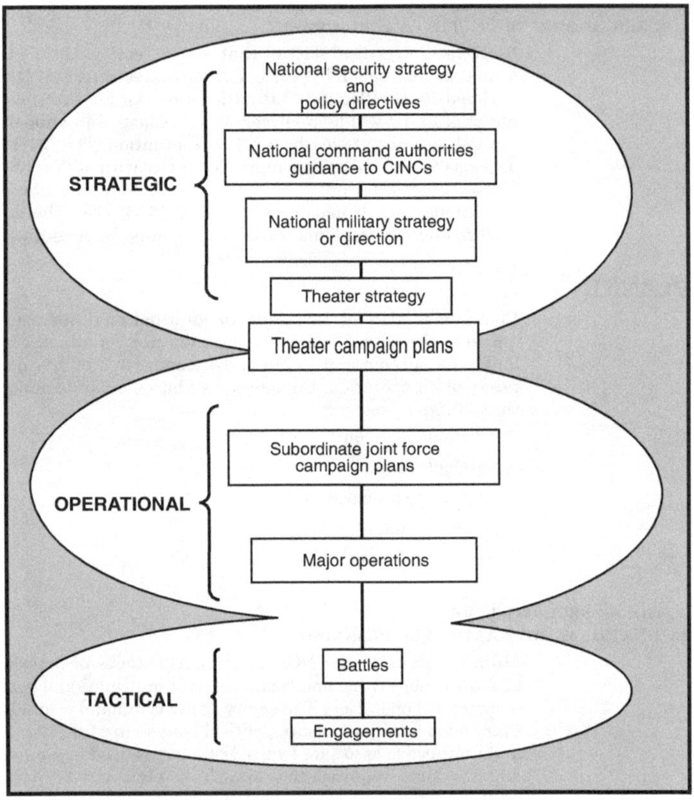

Figure 5-1. Hierarchy of Guidance and Implementing Operations

At the operational level of war, joint and combined operational forces within a theater of operations perform subordinate campaigns and major operations and plan, conduct, and sustain to accomplish the strategic objectives of the unified commander or higher military authority.

The operational level is the vital link between national- and theater-strategic aims and the tactical employment of forces on the battlefield. The focus at this level is on conducting joint operations—the employment of military forces to attain theater-strategic objectives in a theater of war and operational objectives in the theaters of operations through design, organization, and execution of subordinate campaigns and major operations.

Combatant commanders and theater-of-operations commanders usually plan and execute campaigns. Combatant commanders have strategic intents, concepts, and objectives. Service or subordinate joint commanders have operational intents, concepts, and objectives in support of the combatant commanders. The echelon of Army commands varies with the nature of warfare, the strategic objectives, the size and structure of the theater of war, and the number of forces involved. The intended purpose, not the level of command, determines whether an Army unit functions at the operational level. Armies normally design the major ground operations of a subordinate campaign, while corps and divisions fight tactical battles and engagements. A corps commander might also command a joint force land component or a JTF. As a JTF commander, he might plan and execute a campaign that would achieve the theater-strategic objectives of the CINC's theater campaign.

Force Projection Planning

Force projection is the demonstrated ability to rapidly alert, mobilize, deploy, and operate anywhere in the world. This deployment, especially if multinational in character, is critical to overall mission success. Multinational considerations must be considered from the beginning to ensure smooth deployment of forces and the most effective use of lift assets. Every deployment is unique, but all have the same eight phases: mobilization, predeployment activities, deployment, entry operations, operations, war termination and postconflict operations, redeployment, and reconstitution demobilization. Predeployment activities and deployment are covered here.

Multinational operations are often characterized by duplication of effort and unit capabilities, with commensurate shortfalls as well. For example, before the UNPROFOR deployed to the Former Republic of Yugoslavia, each participating nation performed its own engineer reconnaissance of the infrastructure. This resulted in many points being surveyed several times and other areas not at all. The MNF must coordinate and anticipate requirements during this phase to maximize capabilities and efficiency and minimize force structure. Planners must review national military contingents and HN assets and agree on division of labor and who provides what forces.

Limited strategic lift calls for maximizing the efficiency of strategic flow when planners coordinate the kinds and sequence of units deployed. This also requires coordination with the HN so that units do not deploy capabilities already available, such as port operations forces.

In some cases, the US may transport foreign forces into the theater. If so, these forces must be added to the time-phased force and deployment data (TPFDD). LNOs from national contingents either must coordinate directly with the US Joint Movement Center (JMC), if the US is moving the forces, or the MNF headquarters. Procedures may vary with each MNF. The JMC coordinates strategic movements with USTRANSCOM and oversees execution of theater transportation priorities. It plans future operations and monitors overall theater transportation performance. It is normally organized along functional lines. The JMC is the nucleus of an organization that can be expanded in proportion to the size of a joint force. In some operations, the JMC may be a US organization while in others the MNF may have its own JMC. In large operations/campaigns, there is normally both a US and a multinational JMC. In such a case, the multinational JMC (sometimes called the multinational deployment agency [MDA]) is the lead movement agency.

Training and Leader Development Planning

Since units cannot make major changes to doctrine, organization, and materiel, planners should focus on training and leader development in peacetime. Transition planning must begin early in the planning process, because US forces may be involved in a variety of operations. Including NGOs and PVOs in command post exercises (CPXs) and exercises will assist the ultimate transition of responsibilities to another organization.

Training should integrate multinational and interagency play in exercises and simulations. Mobile liaison teams (MLTs) and other designated liaison organizations should be trained, deployed, and exercised. Requests for linguists and participation of reserve component (RC) units that provides qualified linguists would also assist in multinational training.

Leader development is enhanced by—

- Instructing domestic operations in service schools, inviting key civilians to attend service schools, increasing attendance by key military personnel at civilian emergency management courses, and increasing mutual military and civilian participation in exercises and simulations.
- Developing centralized procedures to identify US officers with foreign language experience, that are not part of the FAO program, for assignment to multinational duties when required.
- Emphasizing the use of campaign planning techniques for peace operations and orienting key civilians for early participation in planning.
- Providing liaison teams with references to appropriate regional information and the language, organization, materiel, and doctrine of multinational partners.

Training is the best method of integrating, learning strengths and weaknesses, and forming effective MNFs. Training should be done at all levels of command, and staffs should learn to use METT-T analysis to determine mission assignments. Before deployment, CPX simulations can provide training in staff action tasks and solve problems in the MNF command structure. After deployment and at the unit level, field training exercises should be employed.

A comprehensive training program helps commanders at all levels identify weaknesses in the MNF and helps build cohesion and mutual confidence. Whenever possible, senior officer seminars and conferences should be arranged to develop and/or agree to SOPs and TTPs.

SUPPORT PLANNING

Support of the force must be engineered concurrently with mission clarification and determination of force composition. Commanders analyze the coalition members' capabilities and willingness to support their organic elements and other components of the force. Assembling the optimal array of support assets and establishing support relationships and procedures are essential to achieving the maximum effectiveness of the operational/combat units. This planning must receive emphasis equivalent to the combat planning.

TRANSITION PLANNING

Most future operations will see a transition from US control of an area/ operation to either UN force control, HN military control, or HN civilian control. Such transition must be planned for up front as part of conflict termination and should be an integral part of staff training. The general pattern is the same for all three cases, so the following technique from the 10th Mountain Division is illustrative. The 10th Mountain Division was the ARFOR and overall MNF commander tasked to create a "stable and secure environment" in Haiti before transition to UN control. The UN resolution authorizing the UN force that would assume responsibility from the 10th Mountain Division limited its size to 6000 troops total, a portion of which would be US. The 10th Mountain planners realized that the best way to demonstrate a safe and secure environment was with a force of 6000 US soldiers to show that the UN force could continue to maintain the situation. Planners developed a glide path troop reduction schedule to reach an approximately 6000-soldier strength that was militarily effective and included the US portion of the UN force. This was done in consonance with the UN Mission in Haiti planners to smooth the transition. As UN forces arrived and were prepared for duty, a smooth hand over could occur, one sector at a time. This technique frees up US forces, efficiently transitions to UN forces, and best accomplishes all other aspects of the overall UN mission.

PREPARATION

In order to successfully execute a multinational operation, plans must be transformed into actions that prepare participating nations for what they have pledged to undertake. At a minimum, this preparation ensures that each nation can maintain mission focus, synergistically balance the available capabilities, and clearly organize future transfers of authority.

FOCUSING ON THE MISSION

Political considerations are the single most important factor in multinational operations. The most critical actions for the commander are to remain focused on the assigned mission and understand the reason each national contingent is participating. This determines the very structure of the MNF, since it may cause a splitting of the overall force into components operating under differing political direction.

The MNF commander orchestrates the efforts of a force whose individual armies represent nations that may agree on a common enemy but have widely diverging views of how to prosecute the war or conflict. The commander must recognize that political considerations may force him to choose an acceptable course of action, rather than the optimum (from a purely military view) course of action. He must retain the flexibility to adjust to unforeseen political influences and keep the MNFs focused on the military objective without allowing mission creep to occur.

To overcome differences in doctrine, training, or equipment, leaders may assign selected functions to a smaller group of partners. For example, the MNF commander could assign home defense or police forces the mission of rear area security. The commander may also entrust one member of the MNF with AD, coastal defense, or some special operation, based on the threat force's special capabilities. In fact, some partners might not provide military forces at all but would contribute through the political power provided by their membership in the alliance or coalition.

Balancing Capabilities

The multinational commander must recognize the strengths and differences of the cultures from which these forces come. Decisions on employment, made in consultation with the military leadership, must consider the capabilities of the MNF. The multinational commander must carefully balance the allocation of capabilities. Subordinate commanders may request control of forces that provides capabilities not organic to that nation's forces. The guiding principle is to allocate assets as they are needed, while maintaining a concentration of critical capabilities.

Transferring of Authority

At some point, national units come under the control of the designated multinational commander. This process, known as TOA, must be accomplished as early as possible.

The exact timing of the transfer must be part of the initial negotiations that govern the formation of the coalition. Established alliances generally have established TOA procedures. Where forces are destined for assignment to the MNF and TOA has not taken place, planners must determine where the TOA (and the follow-on integration of units and headquarters) occurs.

One option is to grant TOA to the MNF commander prior to deployment—within each participating country (at a unit's home station). This allows the MNF commander to control unit arrival sequence to best suit operational requirements and facilitates reception area base operations such as feeding and billeting. This option assumes clear political consensus, timely decisions on national participation, and a significant lead time for planning and setting up the MNF headquarters. There are no known historical examples of this option.

A more likely approach is to have TOA at an ISB en route to the operational area. In Uphold Democracy, TOA occurred at an ISB in Puerto Rico. Problems are resolved in a secure area, and forces deployed only when fully ready and in the sequence required by the MNF commander.

The third option is to have TOA occur upon arrival, or soon afterwards, of forces in the operating area. This third option leaves each nation with the responsibility to deploy its contingent and prepare it for operations. It does not allow the MNF commander positive control of deployment into the operating area. It is a less than optimum TOA if immediate combat actions are likely.

Whichever option is chosen, central coordination of deploying forces is preferable. Coordinated planning and execution of the deployment assure that reception operations are not done by repetitive crisis management. Centralized control of force flow provides the best support to the commander's tactical requirements and the best support to the forces and personnel.

Command jurisdiction is the legal position of command by one national authority over the soldiers of another nation. It is important to MNFs which are augmented and supplemented by other national elements. National sensitivities must be carefully considered as well as differing norms of behavior among national militaries and civilian agencies.

Appendix A

Guide for Coalition Operations

While alliances have developed effective methods for integrating forces, most coalitions are created on short notice. As a result, coalition members must develop detailed plans quickly. This coalition planning guide will assist commanders and their staffs during the decision-making process in organizing their efforts to integrate their forces into the overall coalition structure. This guide is not all-inclusive. Additional actions may be required, depending on the coalition's operational mission and/or composition.

MISSION ANALYSIS

1. Determine the desired political objective and end-state condition as defined by the legitimizing political authority.
   - Establish the military role in attaining this political end state. Coordinate the military role with the roles of other federal agencies, NGOs, and PVOs involved.
   - Analyze the mission for clear and attainable military objectives. Ensure the commander's intent supports the desired political end state.
   - Translate these objectives into tasks for subordinate commanders.
2. Select the nation with the greater national interest and/or the best ability to plan for and direct the MNF to lead the effort.
3. Establish the probable cost (in lives, money, resources) of the operation and if the cost is acceptable.
4. Determine which legitimizing political authority defined the desired end state and the relationship between the military force and this political entity.
5. Determine the national end-state criteria of each coalition partner and if it differs from the criteria of the MNF itself.
   - Select alternate courses of action the MNF will follow when a national military element withdraws from the force (that is, actions following decomposition of the force).
   - Determine courses of action to be executed if the sponsoring organization orders withdrawal of MNF in advance of end-state achievement.
   - Determine redeployment and/or withdrawal plans for MNF and how the departure of forces is accomplished under tactical conditions.
   - Decide actions to be taken in the event one member of an MNF executes a unilateral withdrawal from the coalition (whether for political or other reasons).

CRITICAL OPERATING SYSTEMS

1. C^2. Select C^2 structure and establish subordinate relationships (OPCON, TACON, national control retained).
   - Establish a minimum communications suite for each of the coalition members to enable C^2 multinational operations.
   - Identify the C^2 systems support required for the diminishing MNF presence.
2. Intelligence. Determine how intelligence and information are shared so that all coalition partners are treated equally.
3. Logistics. Agree on a logistical support structure identifying capabilities and responsibilities.

CONTROL MECHANISMS

1. Language.
   - Agree on *coalition language* for forcewide use.
   - Determine command level at which each force resorts to its national language and does not use interpreters.
   - Select sufficient interpreters for planning and execution.
2. Liaison.
   - Identify and interview key LNOs.
   - Determine if LNOs will be sent to the various coalition headquarters. Identify their communication requirements. If coalition LNOs are sent to the US headquarters, determine communications support and augmentation.

RULES OF ENGAGEMENT

1. Determine the ROEs, who sets them, and when they are reviewed.
2. Determine how national ROEs affect the MNF organizations and operations.
3. Decide on procedures used by commanders to request a change to the current ROE.

STAFF TASKS

The following tasks are performed by the various staff agencies indicated.

PERSONNEL/G1

1. Identify special skill requirements (unit and/or individual). Establish requirement for RC units and/or individuals. Devise a personnel replacement and rotation scheme.
2. Request language-qualified personnel for augmentation. Request language *survival kits.* Identify available training such as MLTs.
3. Coordinate with G3 and request available augmentation, such as MLTs and joint movement control teams.

INTELLIGENCE/G2

1. Determine how US strategic intelligence is shared among other forces.
2. Provide other national forces intelligence summaries, imagery, and threat assessments approved for dissemination.
3. Assess overall MNF intelligence gathering and dissemination capabilities. Prescribe and disseminate procedures for intelligence sharing and reporting.
4. Determine available linguistic support and/or requirements. Coordinate with the G1.
5. Determine any unique relationships (cultural, historical, other) between MNF partners and the adversary and/or enemy.

OPERATIONS/G3

1. Prepare organization chart for the ARFOR and coalition forces.
2. Prepare capabilities brief/description for ARFOR and coalition forces.
3. Determine effect of national ROEs and objectives on force composition and mission assignment.
4. Determine the current situation.
5. Decide how mission, to include commander's intent, is disseminated and checked for understanding and consistency.
6. Begin planning for transition to UN or other organizations that will take over from the MNF. Contact counterpart planning staffs.
7. Determine logistical situation.
8. Establish language and interpreter requirements.
9. Determine special requirements.
10. Review special customs and courtesies.
11. Provide SOPs that include reporting requirements and procedures.
12. Establish times when units are available.
13. Coordinate ground and air reconnaissance and establish a common map data base.
14. Coordinate staff visits.
15. Coordinate unit commander's visit to higher headquarters.
16. When redeploying or moving, determine what sector forces remain to support the coalition and how long they are required.
17. If required, determine if foreign forces will interface with the joint movement control center.
18. Determine training required before deployment and prepare for continued training, including multinational, after deployment.

LOGISTICS/G4

1. Determine composition of multinational logistics command and/or element.
2. Divide responsibilities between multinational, national, and HN.
3. Determine how each class of supply will be handled.
4. Transportation.
    - Determine the combined transportation command structure.
    - Determine the coalition reception, staging, onward movement, and integration process.
    - Determine if coalition forces have tactical rotary- and fixed-wing assets for intratheater supply.
    - Determine who supplies transportation supply throughput for coalition forces from the JTF logistics center.
    - Determine if coalition forces have transportation assets for the movement of troops.
5. Supply.
    - Identify the coalition forces' capabilities to receive, store, and issue dry cargo, fuel, and water to include water production and/or purification capability.
    - Determine if coalition forces have the means to communicate requirements to the JTF logistics center or coalition logistics management center.
    - Identify availability of materiel handling equipment within the coalition.
    - Identify coalition air and sea LOCs.
    - Determine if coalition forces have a basic load of ammunition and their Class V procedures.
    - Identify the coalition force's special requirements (tents, cots, reverse osmosis water purification units, laundry, latrines, batteries, etc.).
    - Determine map requirements for coalition forces.
6. Mortuary Affairs. Determine capabilities of the coalition forces.
7. Maintenance.
    - Determine if coalition forces have maintenance support.
    - Determine if coalition forces have the means to order and receive Class IX.
    - Determine if coalition forces have wreckers, stake and platform trailers, or HETs.
    - Determine if coalition forces have communications repair facilities.

CIVIL AFFAIRS/G5

1. Establish and coordinate operations of the CMOC.
2. Identify the NGOs and PVOs; determine their intentions and future plans; identify the key personnel in sector.

INFORMATION MANAGEMENT/G6

1. Identify the command relationships, the location of the headquarters each unit reports to, and the type of service required (tactical satellite, telephone, facsimile, AM, FM).
2. Articulate in the initiating directive that the multinational command channels are for the execution of military operations and national channels are for reporting status and requesting support.
3. Establish a means and a plan to provide a common tactical picture to all forces.
4. Provide coalition partners with a lesser C^2 capability interpreters (if necessary), operators, and maintainers to enable interaction with the commander and other coalition members.
5. Determine the frequency requirements and planning ranges for equipment.
6. Consider the terrain while planning for the C^2 network.
7. Request frequencies from the JTF J6 or coalition communications coordinator.
8. Provide for common data bases.

ENGINEER

1. Identify terrain visualization requirements.
2. Determine types and capabilities of engineer units for the coalition and other services.
3. Determine the facility support requirements from the MNF and its supported units (that is, latrines, base camp construction).
4. Determine the condition of and requirements for infrastructure in the AO (roads, airfields, ports, power generation facilities).
5. Identify the availability and type of engineer resources in the operating area.
6. Determine real estate support requirements.
7. Identify humanitarian and nation assistance engineering requirements.

AVIATION

1. Determine the aviation assets, capabilities, and requirements of the coalition force.
2. Identify the aviation logistics capabilities (fuel, spare parts) of the coalition force.
3. Identify current and projected requirements for an air LOC.
4. Determine aviation support required from US forces.
5. Identify the intended base of operations.

PUBLIC AFFAIRS

1. Determine PA coverage plan.
2. Coordinate with other national PA officers or equivalents.
3. Identify the procedures, if any, for a security review of media material.
4. Establish a plan for handling publicity, news correspondents, and journalists.
5. Determine the coalition senior leaders and their biographical backgrounds. Arrange face-to-face meetings.
6. Identify the CMOC and the participants.
7. Identify the senior operation spokesperson for the MNF.
8. Establish a combined information center.

FINANCIAL MANAGEMENT

1. Finance Operations.
   - Determine if a weapons bounty program is needed. Prepare finance to support weapons bounties, claims.
   - Put arrangements and/or agreements in place to provide or receive multinational support to the local procurement process. For example, funding paying agents for other services or cashing checks to provide cash funding for finance elements of other nations.
   - Determine limitations on the amounts of cash payments (including check cashing) that soldiers may receive in the AO. Determine who imposes limitations—the MNF commander or the HNs (SOFA).
   - Establish requirement for US finance forces to provide other currencies for the operation and currency exchange.
   - Determine if finance will provide currency exchange for personnel of other nations and/or provide currency to their finance representatives and will US forces receive such support.
   - Decide if multinational (contracted) subsistence support to the force will affect entitlements to US soldiers.
2. Resource Management.
   - Provide advice during operational planning and take action to arrange funding authority for possible courses of action.
   - Provide tracking and cost-capturing guidance for MNF agreements on logistics support from or to other nations.
   - Place fund citations and fund sources in place to allow US forces to track expenditures properly.
   - Ensure limits on funding authority are understood by operating forces.
   - Advise commanders as to whether or not expenditures and costs incurred are approved for reimbursement.

LEGAL

1. Review and complete SOFA and ROE.
2. Brief commander/G4/comptroller on legal aspects of supporting foreign forces.
3. Determine if all soldiers were briefed on the ROE prior to assumption of operational control by the coalition or deployment to the operational area.
4. Establish a reporting procedure for all compensable events, accidents, injuries to locals, deaths, or property damage for potential claims.
5. Maintain a record of the logistics support provided by the US force to coalition forces and support the US force receives from the coalition.
6. Report all serious incidents involving HN and/or coalition forces and the US force.
7. Determine if there are any cross-servicing agreements between the US force and other coalition, UN, or HN elements.
8. Review the OPLAN to ensure compliance with US and international law.

SURGEON

1. Determine the capabilities for MEDEVAC and whether the US force will be supported by, or required to support, other coalition elements.
2. Identify the MEDEVAC assets (air and ground).
3. Identify the source of Class VIII supply and payment options.
4. Identify procedures for tracking patients and coordination requirements for return-to-duty transportation.
5. Determine if forces have organic Echelon I, II, or III combat health support. If not, what level will US forces provide.
6. Determine the airevac procedures for US and coalition force personnel (evacuated to the continental US or their home country).
7. Determine the policy and procedures for US medical personnel to use coalition forces Echelon II through V medical treatment facilities to provide medical treatment for US and coalition forces.

Appendix B

Liaison Personnel

Liaison, in the military sense, refers to contact or intercommunication maintained between parts of an armed force to ensure mutual understanding and unity of purpose and action. It is often aided by the exchange of personnel to facilitate an exchange of information. Liaison is therefore a vital function that assumes increased importance in the conduct of multinational operations.

RANK

An LNO must have sufficient rank and authority to speak for his commander. Many armies are extremely rank conscious and will not deal with an LNO of lesser rank. As a rule, the LNO should be of equal rank to the chief of staff or executive officer.

CHARACTERISTICS

LNOs must be soldiers of quality and experience, able to represent their commander's intentions and report objectively to him. On occasion, they may make decisions on behalf of the commanders they represent. LNOs do not merely convey decisions after the fact. The commander consults his LNO during planning for advice on how to employ his country's forces and how to convey the orders.

Of prime importance is the personality of the LNO. He must inspire and encourage friendliness, confidence, and comradeship with the officers and members of the unit from which he seeks information. He must establish himself as an integral working part of the headquarters. He must always respect the military procedures and customs of the unit to which assigned and observe the established channels of staff and command functions. If he needs further verification or more detail on a subject, the LNO must request them diplomatically. In realizing that he personally represents his commander, the LNO must react accordingly. If problems are caused by personalities, the LNO's commander must iron them out or change liaison assignments. Cooperation is the objective of all parties; personality problems should not interfere. The LNO must be thoroughly familiar with command and staff procedures, operations, and Army organization, to include a detailed knowledge of the table of equipment of his parent unit and that of the unit to which he is sent. To get the most from information obtained at the visited headquarters, the LNO must keep up to date on the situation, which may prove to be conducive to obtaining more information from the visited headquarters. The old adage "As you give, so shall you receive" should be his guide in this respect.

Because they are not impressed with the value of their assignments, many LNOs see their role as nothing more than glorified messengers. This lack of enthusiasm is furthered by the receiving headquarters' reception of LNOs, which can usually best be described as indifferent. The visiting liaison personnel are told where to wait for their information. A situation map may be provided, and they may only receive information deemed essential by the visited headquarters. Such attitudes and procedures stifle, somewhat, the personal initiative and abilities of liaison personnel. It is the LNO's responsibility to improve this situation.

Very often the professional LNO can help prevent last-minute changes by participating in the planning process of the headquarters to which he is attached. Allied or coalition headquarters may be reluctant to pass preliminary plans that are subject to change to other headquarters. The presence of an LNO in the planning process can speed coordination of emerging plans among headquarters. The LNO must therefore understand the US commander's overarching intent and strive to keep up to date on the situation. LNOs should attend planning conferences, to make notes of coming operations, and to report them. If any critical change or addition is suggested from his own headquarters, the LNO should obtain these in time to prevent amendment or revision of orders by either headquarters.

The dispatching headquarters has certain responsibilities for its liaison teams. The commander and staff sections must know the identities of their liaison team. Frequently, subordinate units are tasked to provide liaison teams whose identities and qualifications are unknown to the headquarters relying on their abilities and information. This *acquaintance process* can be accomplished by a face-to-face meeting between the responsible headquarters personnel and the liaison team. During this meeting the liaison team should receive a briefing on its mission, the situation, and the unit to which it is being sent. In addition, the team should devote considerable attention to reporting procedures; providing secure, reliable communications and signal operating instructions. Radio is normally the quickest and easiest means of communication. However, give consideration to the fact that the opening and closing of liaison radio nets are positive intelligence indicators. Historically, these nets have proven to be a lucrative source of information for an enemy.

TRAINING

A firm procedure for the training of liaison personnel is difficult to establish—so much depends on the personality traits and military knowledge of the individuals. As a minimum, LNOs must be thoroughly grounded in their own unit's command and staff procedures, doctrine, organization, capabilities, and communications equipment and procedures. LNOs must have the same information on the allied headquarters to which they will go. They must also know the purpose of liaison, the liaison system in effect, reporting methods, records and documents they are to maintain, their mission, and how they should conduct themselves. Finally, the liaison party, section, or mission must train and exercise as a group. A liaison party that is thrown together at the last minute is less effective, and operations may not allow a suitable break-in period. Group training provides a backup capability for the liaison team, allowing it to function even if key personnel are lost and not readily replaced.

SELECTION

Ensure maximum effectiveness of liaison missions by selecting an officer who—

- Has the confidence of the sending unit commander.
- Is favorably known, either personally or by reputation, by the commander and staff of the receiving unit.
- Has a sound and comprehensive knowledge of military operations.
- Is tactful.
- Has experience or training as an LNO.
- Has the necessary language expertise (if required).

DUTIES

Prior to departure, the LNO should—

- Become familiar with the situation of his own unit and that of the receiving unit.
- Clearly understand the mission.
- Ensure that arrangements for communications (signal and transport) are adequate.
- Obtain credentials in writing unless obviously unnecessary.

Upon arrival at the receiving headquarters, the LNO should—

- Report promptly to the commander, state his mission, and show any written directive or credentials.
- Offer assistance to the commander, if appropriate.
- Arrange to obtain information required by the mission.
- Familiarize himself with the situation of the unit.
- Establish communications with his parent headquarters.

During his liaison tour, the LNO should—

- Further harmonious cooperation between the two headquarters.
- Accomplish the mission without interfering with the operations of the headquarters.
- Keep informed of the situation of his own unit and make any information available to the commander and staff of his receiving unit.
- Keep an appropriate record of reports.
- Report on those matters within the scope of his mission.
- Advise the visited unit commander of the contents of reports to be sent to his own headquarters.
- Report promptly to the parent headquarters if unable to accomplish the liaison mission.
- Report departure to the visited unit commander on the completion of his mission.
- Make note of personality traits and idiosyncrasies of key commanders and staff officers of the headquarters/units visited. Such information should remain CONFIDENTIAL.

On return to his own headquarters, the LNO should—

- Report on his mission.
- Promptly transmit any requests of the commander from whose headquarters he has just returned.
- Report on key personalities and general operations of the headquarters from which he has just returned.

The sending headquarters should—

- Give the LNO definite and detailed instructions, in writing if appropriate, as to the liaison mission.
- Inform the LNO of the commander's plans, especially as they affect the unit to which he is to be sent.
- Ensure that adequate facilities are available for communications between the LNO and the sending headquarters.
- Brief the LNO in detail about the type of unit to which he is being sent and key personalities he will encounter.

Appendix C

Augmentation

Multinational operations normally require more C^2 capability than the unit or controlling headquarters has available. Force projection also calls for more modularity to allow tailoring of forces. Given these facts, the Army created several organizations that the commander can call upon for assistance when conducting multinational operations. Other organizations exist within various regions that local units may be more familiar with, but the organizations and teams discussed here provide a sample that planners should consider in their planning and training. Although not officially responsible to do so, the Defense Language Institute produces survival kits—normally booklets or cards with basic phrases (stop, lie down)—for deploying forces. With sufficient lead time, the Defense Language Institute can produce a wide range of language-related products.

MOBILE LIAISON TEAM

The Training and Doctrine Command has codified the MLT concept. The MLT consists of approximately 23 functional subject-matter experts to assist joint, multinational, or Army-level commands in joint, multinational, and interagency environments. Six MLTs are available for deployment with one MLT remaining in alert status for rapid response to requirements. MLTs are capable of providing advice and assistance to supported units and ensuring rapid and accurate coordination between headquarters. MLTs have organic transportation, communications, and language capabilities. Language capabilities appropriate for all theaters come from the Defense Language Institute. The Power PAC Company provides organic communications capability for all MLTs.

HOST NATION COORDINATION TEAM

This team consists of a 5-person coordination cell capable of managing and coordinating available HNS, resourcing the contingency force, obtaining local HNS resources through local purchase or contracting actions, and coordinating with financial and legal advisors to ensure proper execution of HNS contracts, as required. The team requires administrative and logistical support. This team is designed for contingency operations only and is intended to be the last team off the first plane in undeveloped areas where no prearranged HNS exists.

LOGISTICS SUPPORT

The US Army Materiel Command's (USAMC) logistics support element (LSE) and the Logistics Civil Augmentation Program (LOGCAP) are nontraditional means of performing a variety of missions associated with logistics support of US and multinational forces involved in military operations.

Logistics Support Element

The LSE is a deployable multifunctional unit that works for the Army component commander and is designed to link the strategic and operational levels of logistics with the tactical units. The LSE is comprised of modular subunits that are called forward by the supported commander as required. The LSE provides C^2 for forward elements of the strategic base which are primarily DOD civilians and contractors. With augmentation, the LSE can provide support to multinational forces. Its capabilities include depot maintenance, calibration of test equipment, ammunition surveillance, release of pre-positioned strategic stocks, and battle-damage assessment.

Logistics Civil Augmentation Program

US armed forces use of contractors to provide supplies and services during both peacetime and contingencies dates back to the Revolutionary War. Today, a program exists to preplan for the effective use of civilian contractors in wartime and other contingencies to augment US forces and support DOD missions. The US Army's Deputy Chief of Staff for Logistics is the proponent for LOGCAP and USAMC administers the contract.

LOGCAP is a Department of the Army capstone program that includes all preplanned logistics, engineering, and construction-oriented contingency contracts actually awarded and peacetime contracts that include contingency clauses. Preplanned weapon system sustainment contracts, ASCC contingency contracts, and the USAMC support contract are prime examples of augmentation contracts that fall under the auspices of the LOGCAP capstone program.

The fundamental goals of LOGCAP are—

- To plan during peacetime for the effective use of contractor support in a contingency or crisis.
- To leverage global and regional corporate resources as facility and logistics services support multipliers.
- To provide an alternative augmentation capability to meet facility and logistics services shortfalls.
- To provide a quick reaction to contingency or crisis requirements.

LOGCAP provides another augmentation support alternative by capitalizing on the civilian sector both inside and outside the continental US. To meet identified logistics, engineering, and construction-services requirement, the CINC and ASCC consider using the following sources (normally in this order), based on availability and other factors—

- Organic support (active component and RC, other DOD agencies).
- Coalition forces and HNS.
- LOGCAP.

During a contingency, the CINC and/or ASCC normally establishes an acquisition review board to determine the optimum means for satisfying CS and CSS requirements based on criticality, timeliness, quality, administration effort, and cost. At a minimum, the following agencies should provide representatives to this board:

- Assistant chief of staff, G4 (logistics), serves as the chairman.
- LSE commander, responsible for team LOGCAP.
- Theater support command.
- US liaison office to MOD.
- HN.
- US Allied Command Europe (USACE).
- Resource management.
- Regional contracting agency.
- Defense Logistics Agency.

Team LOGCAP's mission is to provide a central management structure and conduit of information for ensuring the smooth execution of LOGCAP requirements. The LSE commander provides coordinating authority over team LOGCAP; this element is a selectively manned, equipped, and trained team. It is prepared to deploy worldwide in support of any contingency requiring LOGCAP capabilities. The team can—

- Advise the requiring activity on LOGCAP capabilities.
- Integrate LOGCAP augmentation capabilities into the deployed force structure to meet METT-T requirements.
- Assist the customer in articulating approved logistics and construction requirements to the contractor.
- Ensure compliance and facilitate the teaming of the customer and contractor to accomplish the mission.

The LOGCAP has tremendous capability and potential. It is, however, an asset that can be misused and abused. Prior planning before its use and control during execution of contingency support ensures that it remains a valuable CS and CSS force multiplier.

Engineer Support

The challenge for engineer planners and executors is to achieve the optimal mix of contractor and military engineer-unit capabilities. Construction contractors are best suited for the longer-duration, heavy construction work in stabilized environments. In turn, contractors leverage local resources (labor and materiel) to minimize costs and impact on intratheater lift and port facilities. The contractor's presence contributes significantly to local-area political and economical stabilization and thereby reduces the need for the presence of US security forces. US commanders in theater must recognize the need for military oversight of contract and contractor activities in the areas of project management, financial management, quality assurance, and audit.

During force-projection operations, extensive contracted civilian-engineer capabilities will probably be available only after D+30 due to mobilization and deployment time lines. Civilian-engineer contracting may be available sooner when deliberately and properly planned for during permissive entry conditions. As Army engineers deploy into the theater, they may be joined by contracted civilian engineers. The Army engineer staff should consider the following when coordinating engineer plans and operations with contracted civilian engineers:

- Requesting the latest engineer intelligence data from any contractors working in the theater to help identify force-projection theater Army engineer requirements and enemy engineer capabilities. (Requirements include availability of real estate, construction materials, and facilities; data on threat mines and obstacles and soils; and construction support from the HN.)
- Establishing engineer staff links between the Army and contracted civilian-engineer staff through the JTF, USACE, or naval facility and the theater engineer staff and headquarters.
- Providing necessary Army engineer LNO support.
- Developing time lines that quickly phase in contracted civilian-engineer capabilities to relieve a deployed Army.

Appendix D

Host Nation Support

Host nation support (HNS) is the civil and military assistance provided by an HN to the forces located in or transiting through that HN's territory. HNS will play a critical role in future operations; unfortunately, outside of the logistics community, HNS is neither well known nor understood across the Army. One Force XXI characteristic is force tailoring in order to deploy only the minimum number of forces into an operating area. Efficient use of available HNS can greatly aid a tailored force package and augment the deployed sustainment force structure. This appendix provides general information on HNS and offers sample considerations for planners to use as a basis for work in a multinational setting.

TYPES OF SUPPORT

The two types of support are planned and *ad hoc*.

PLANNED

Planned HNS is support that has been identified and requested during peacetime for provision in time of crisis or war. Based on negotiations between the HN and the sending nations, agreements are developed to obtain assured HNS for all programmed force requirements. The multinational logistics coordinator must be aware of all agreements between or among participating nations.

Requirements planning should be for support of generic units operating within a bilateral scenario. Identification of the absolute needs of each specific unit or developing data bases of HN capabilities is ideal. However, the absolute minimum planning product should contain accepted procedures which will be used to establish the HNS support of force requirements when sufficient information is available.

AD HOC

Ad hoc HNS is support requested during crisis or war that was not identified, planned, or agreed upon during peacetime. HN assets are not unlimited, and the HN will probably give its own population and forces priority of support. Therefore, *ad hoc* requests for HNS may neither be fully fulfilled nor provided to support valid requirements.

LOGISTICS

An operating areas' infrastructure is a key source of logistics. Common areas of HNS are facilities, transportation, services and civilian supplies, labor, construction equipment and materials, and base operations support functions. Three factors influence the use of HNS in an operation—

- Capability, dependability, and willingness of HN to provide assistance.
- Risk associated with a dependence on HNS.
- The effect on security of US forces.

HNS may come from HN governments, civilians, military units, or facilities and may be broken down by function or area support.

AGREEMENTS

HNS agreements can be numerous and complex and require deliberate and continuous coordination among national contingents. This should prevent inadvertent conflicts and price escalations due to demand outpacing supply. Bilateral agreements entered into by members of the coalition must be revealed to other members of the coalition (typically, however, the US does not reveal financial-related information). Maximizing HNS can result in shorter lead-time support to the forces. It can also benefit the local area economy. The HN infrastructure must not be jeopardized by requiring too large a portion of a given commodity or service. Prudent use of HNS must allow the mission to be performed even if the HN provider fails to deliver the agreed-upon materiel.

The authority of the US to negotiate and conclude agreements with foreign governments is assigned by US Code to the DOS within the framework of implementing government regulations. However, negotiations and conclusion of certain aspects of HNS agreements are delegated to DOD and ultimately to unified commanders. The process described here reflects this delegation of authority and the procedures the US established to conclude these agreements. Because some nations prefer to exercise their sovereignty in a different manner, local variations are sometimes required, but they usually conform to the principles defined within this overall framework.

Figure D-1 depicts the process for entering into and concluding negotiations for HNS agreements between the US and a foreign government. The process usually starts with the *UCP* promulgated by the NCA. The *UCP*—

- Establishes the unified commands.
- Assigns them geographic responsibility.
- Assigns primary tasks to be carried out by the command.
- Defines the command relationships within the theater and between the theater and other US commands.
- Provides guidance on the exercise of combatant command authority.

For example, within the European theater, the geographic responsibility of the US command is greater than the area covered by NATO, and the command has relationships with other non-NATO nations. The *UCP* establishes implicitly the need to enter into HNS agreements with these nations and those within the alliance. Additionally, the defense planning guidance (DPG) and the Joint Strategic Capabilities Plan (JSCP) provide more detailed direction regarding specific planning responsibilities and the forces available to execute the plans. The OPLANs and CONPLANs developed by the unified commander, his staff, and the components provide the more detailed information on how US forces might be employed to accomplish the tasks outlined in the *UCP* and other guidance documents.

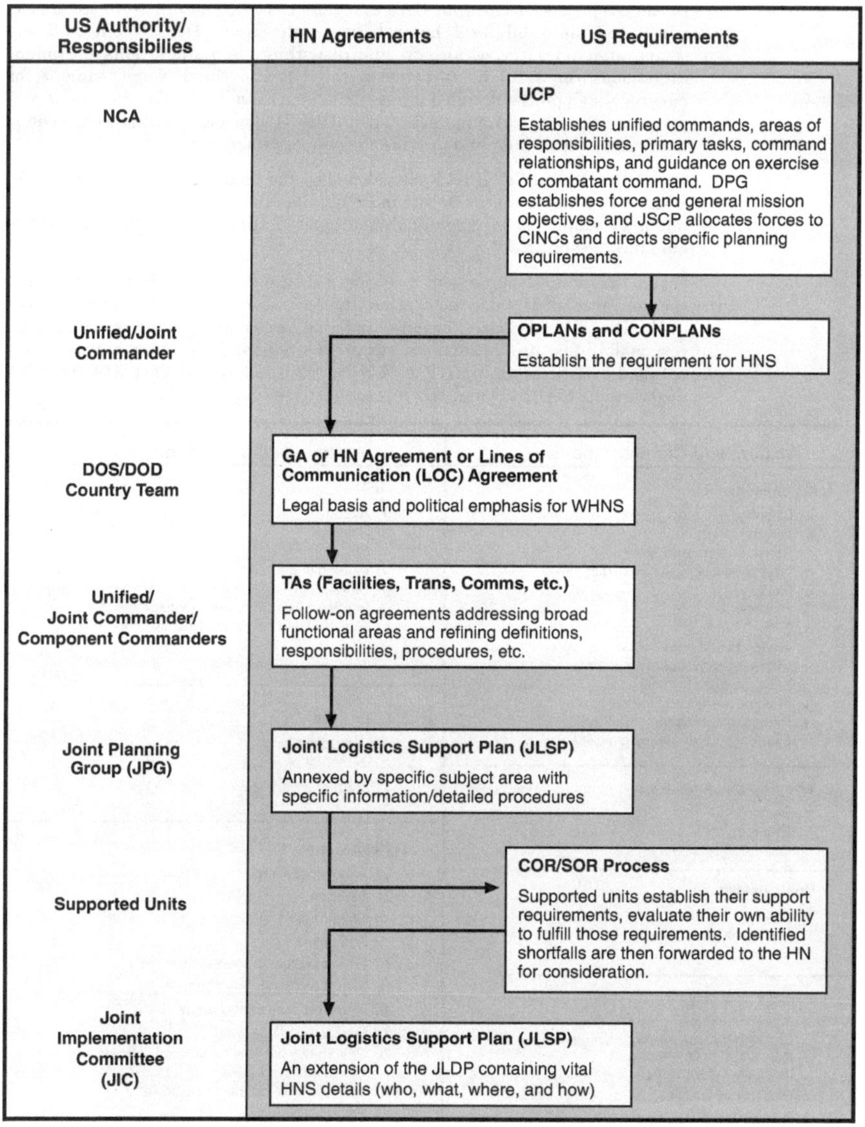

Figure D-1. US Host Nation Support Agreement Negotiation Process

In the US process, the initial HNS agreement is the *general agreement* that is negotiated on a bilateral basis between the State Department and the comparable ministry of foreign affairs within the nation. This document establishes the legal basis between the US and the HN government for moving ahead on the detailed agreements to achieve HNS. The document also provides the political emphasis within the HN to ensure that the various national agencies plan and provide the support when required.

The OPLANs and CONPLANs developed by the unified commander, his staff, and the components provide the more detailed information on how US forces might be employed to accomplish the tasks outlined in the *UCP* and other guidance documents.

When the general agreement is concluded, the commands begin developing the next level of HN documentation, the *technical agreement* or *arrangement* (TA). The TA addresses broad functional areas and includes definitions, responsibilities, procedures, etc., to provide guidance to the components for detailed HNS planning. The US process explicitly restricts the TA negotiations to the US and HN representatives (Figure D-2).

Item and Considerations	Item and Considerations
1. Accommodations a. Billeting b. Offices c. Stores and warehouses d. Workshops, vehicle parks, gun parks e. Medical f. Hardstands g. Fuel h. Weapons and ammunition i. Transportation, including aircraft j. Firing ranges k. Training areas and facilities l. Recreation areas and facilities m. Laundry and dry cleaning facilities	**5. Fuel** a. Aircraft b. Vehicles c. Ships d. Method of delivery e. Storage f. Interoperability of refueling equipment g. Common use of refueling installations
2. Weapons and Ammunition a. Security b. Storage c. Collection or delivery	**6. Local Labor** a. Method of hiring b. Method of payment c. Administration
3. Communications a. Local b. International c. Security	**7. Maintenance** a. Accommodations b. Vehicles c. Ships d. Equipment e. Roads (including snow clearance) f. Fixed- and rotary-wing aircraft g. Provision of assembly areas h. Damage control i. Emergency facilities for visitors' vehicles and equipment j. Evacuation of disabled vehicles and equipment
4. Finance. Payment for – a. Accommodations, supplies, communications, equipment, local labor, maintenance, medical, and movement facilities b. Emergency facilities c. Personal facilities	

Figure D-2. A General List of Items Which May Be Included In A Host Nation Support Agreement

Item and Considerations	Item and Considerations
8. Medical a. Normal facilities b. Emergency facilities c. Reciprocal national health agreements d. Evacuation of casualties	**11. Supplies and equipment** (other than ammunition, fuel or rations) a. Common use items **12. Translation** a. Interpreters/Language Specialists b. Translation of documents
9. Movement a. Airheads (1) Alternates (2) Facilities (3) Ships (4) Equipment (5) Refueling b. Ports (1) Alternates (2) Draft (3) Bunkering/Fueling (4) Repair c. Road and rail movement (1) Personnel (2) Equipment (3) Security (4) Traffic control d. Pipeline movement	**13. Transportation equipment** a. Use of host nation military vehicles, equipment, ships, aircraft b. Locally hired vehicles and equipment, ships, aircraft c. Policy on drivers and handlers for b above **14. Water** a. Production/purification capability (1) Municipal (2) Other water treatment systems b. Distribution capability (1) Trucks (2) Pipeline (3) Hoseline c. Storage capability d. Receipt and issue capability e. Available water sources (1) Wells (2) Surface (3) Subsurface f. Host nation water quality standards
10. Rations a. Fresh b. Pack c. Potable water	**Note:** While each multinational arrangement is unique, NATO arrangements provide a baseline to use and modify as necessary.

Figure D-2. A General List of Items Which May Be Included In A Host Nation Support Agreement (continued)

The next step, *level of agreement*, is conducted by a joint planning commission or group (JPC/G), co-chaired by the senior officials of the US unified command and the HN ministry of defense (MOD). This group codifies the policy and guidance needed to conduct joint bilateral HNS planning into a document, annexed by specific subject area, and known as the Joint Logistics Support Plan (JLSP). The JLSP guidance identifies the type of support the HN is able and willing to provide, and the conditions under which the support is made available. The next step in planning is for the sending nation to develop and submit to the host nation their *concept of requirements* (CORs).The COR is a statement of functional HNS requirements that the components need to support deploying US forces. The US process explicitly restricts the TA negotiations to the US and HN representatives (Figure D-2). Once the COR is submitted and approved (to the extent possible by the HN), the statement of requirements (SOR) is submitted.The SOR identifies in detail the type and amount of HNS required by the unit to accomplish its tasks. The requirements may be time-phased and incorporate facilities, services, and supplies at several separate locations. Generally, however, it should be expressed as a quantifiable work load that the HN can plan to allocate resources against during execution of US operations. Figure D-3 depicts a sample of the type of requirements contained in the COR/SOR.

Host Nation Support Considerations	Multinational Logistics Support Plan	Multinational Lines of Communication Plan	Multinational Facilities Plan
Control and Coordination			
Border crossing	X	X	
Command arrangements	X	X	X
Currency exchange	X	X	X
Customs	X	X	
Damages	X	X	X
Finance	X	X	X
Liaison	X	X	X
Movement control	X	X	X
Press and visitors	X	X	X
Status of forces	X	X	X
Facilities			
Airfields		X	X
Headquarters and communications	X	X	X
Holding areas	X	X	X
Medical	X	X	X
Seaports	X	X	X
Staging areas	X	X	X
Storage areas	X	X	X
Training areas	X	X	X
Services			
Communication	X	X	X
Construction	X	X	X
Medical evacuation	X	X	X
Maintenance	X	X	X
Health service support	X	X	X
Security	X	X	X
Staging	X	X	X
Storage	X	X	X
Transportation	X	X	
Supplies			
Rations	X	X	X
Ammunition	X	X	X
Fuel	X	X	X
Medical	X	X	X
Other	X	X	X
Personnel			
Civilian personnel and labor force	X	X	X
Military personnel	X	X	X
Enemy prisoners	X	X	
Internees	X	X	
Noncombatants	X	X	

Figure D-3. Host Nation Support Considerations and Documentation

The CORs/SORs are provided to the joint implementation committee (JIC), which is co-chaired by a representative of the sending nation and the appropriate HN territorial command. This committee establishes the detailed procedures for US commands to obtain SOR-related HNS from the nation. These procedures become the Joint Implementation Plan (JIP).

Figure D-4 depicts the process required to modify/renegotiate an existing HNS agreement. The process can be initiated by either country and normally modifications are required when the political or military environment causes changes to national guidance and/or OPLANs. Since an HNS agreement already exists, including existing support and implementation plans, the process flow is similar and primarily involves evaluation of existing documents for sufficiency/accuracy and modifications made as necessary.

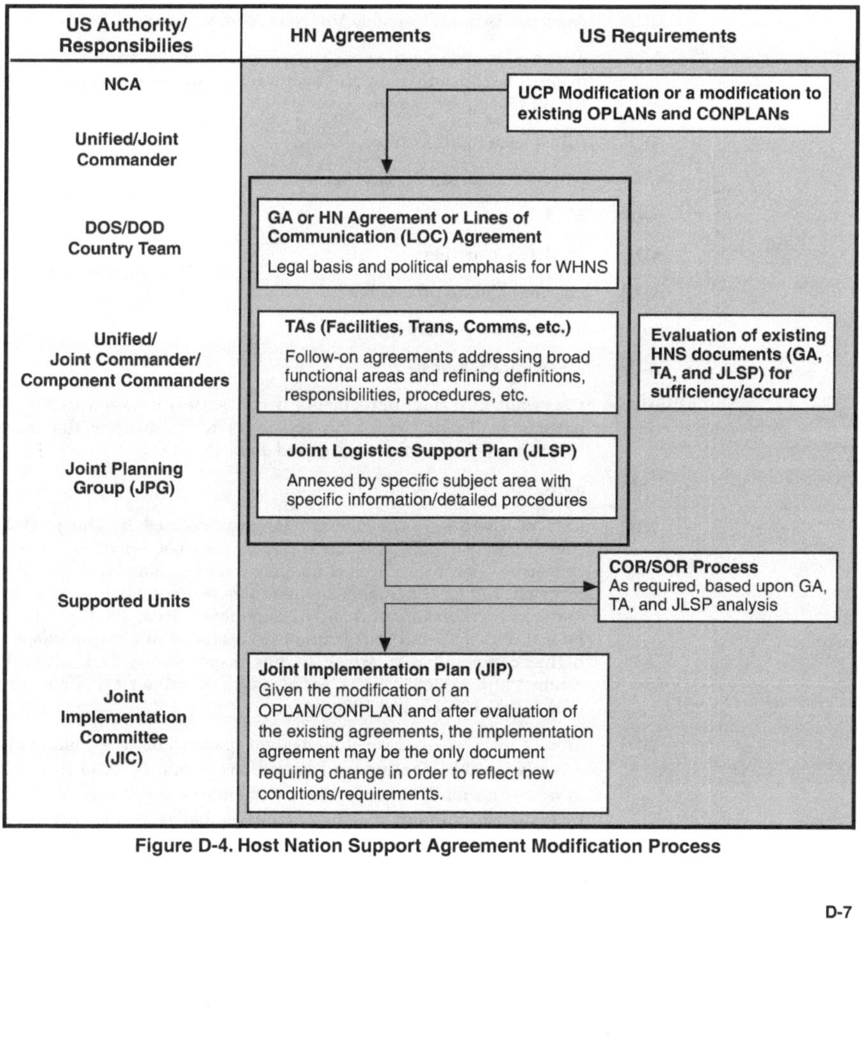

Figure D-4. Host Nation Support Agreement Modification Process

Glossary

AADC area air defense commander—the overall responsibility for air defense assigned to a single commander within a unified command, subordinate united command, or joint task force; this is normally the component commander with the preponderance of air defense capabilities and the command, control, and communications capabilities to plan and execute integrated air defense operations; other components involved provide representation, as appropriate, to the area air defense commander's headquarters

AAMDC Army Air and Missile Defense Command (Provisional)

ABCA America, Britain, Canada, Australia Armies Standardization Program

ACA airspace control authority—the commander designated to assume overall responsibility for the operation of airspace control system in the airspace control area

ACE Allied Command Europe

ACSA Acquisition Cross-Service Agreements

AD air defense

ADA air defense artillery

ADP automated data processing

airevac aeromedical evacuation

AJP allied joint publication

alliance the result of formal agreements (i.e., treaties) between two or more nations for broad, long-term objectives which further the common interests of the members (proposed Joint Pub 5-0)

AM amplitude modulation

AO area of operation—the operational area defined by the joint force commander for land and naval forces; does not typically encompass the entire operational area of the joint force commander, but should be large enough for component commanders to accomplish their missions and protect their forces; Army—geographical area, usually defined by lateral, forward, and rear boundaries assigned to a commander, by a higher commander, in which he has responsibility and authority to conduct military operations; *see also* area of influence and battlespace (FM 100-5, 100-15, and 100-20)

AOI area of interest—a three-dimensional space wherein a commander is directly capable of influencing operations by maneuver or fire support systems normally under the commander's command or control; includes both organic and supporting combat power, to include joint, multinational, or interagency assets (Joint Pub 1-02)

AOR area of responsibility—the geographical area associated with a combatant command within which a combatant commander has authority to plan and conduct operations; in naval usage, a predefined area of enemy terrain for which supporting ships are responsible for covering by fire on known targets or targets of opportunity and by observation

ARFOR Army force

ARRC allied rapid reaction corps

ASCC Army service component commander—the principal advisor to the commander in chief for supporting and employing Army forces in theater and forces outside the theater tasked to support theater operations; this command consists of those Army individuals, units, detachments, organizations, and installations; responsible for all command aspects of Army forces, to include logistics within the unified command

ATACM Army Tactical Missile System

battlespace the conceptual physical volume in which the commander seeks to dominate the enemy; it expands and contracts in relation to the commander's ability to acquire and engage the enemy or can change as the commander's vision of the battlefield changes; encompasses three dimensions and is influenced by the operational dimensions of time, tempo, depth, and synchronization; is not assigned by a higher commander nor constrained by assigned boundaries

C^2 command and control—the exercise of authority and direction by a properly designated commander over assigned forces in the accomplishment of the mission; functions performed through an arrangement of personnel, equipment, communication, facilities, and procedures employed by a commander in planning, directing, coordinating, and controlling forces and operations in the accomplishment of the mission (Joint Pub 1-02)

C^2W command and control warfare

C^3I command, control, communications, and intelligence

C^3IC coalition coordination, communications, and integration center—established in the absence of a single commander or in-place alliance to coordinate multinational activities; provides unanimity of effort among coalition ground forces without the benefits of unity of command (FM 100-7)

C^4I command, control, communications, computers, and intelligence

CA civil affairs

CAS close air support

CFC Combined Forces Command (Korea)

CI	civilian internees
CINC	commander in chief
CJTF	commander, joint task force
CMO	civil-military operations
CMOC	civil-military operations center
coalition	an *ad hoc* arrangement between two or more nations for a common action, for instance, the coalition that defeated Iraqi aggression against Kuwait in the Gulf War, 1990-1991 (proposed Joint Pub 5-0)
coalition action	multinational action outside the bounds of established alliances, usually for single occasions or longer cooperation in a narrow sector of common interest; *see also* combined action (proposed Joint Pub 5-0)
coalition force	a force composed of military elements of nations that have formed a temporary alliance for some specific purpose (Joint Pub 1-02)
COCOM	combatant command (command authority)
combatant command (command authority)	nontransferable command authority established by Title 10, *United States Code*, Section 164, exercised only by commanders of unified or specified combatant commands unless otherwise directed by the President or the Secretary of Defense; the authority of a combatant commander to perform those functions of command over assigned forces involving organizing and employing commands and forces, assigning tasks, designating objectives, and giving authoritative direction over all aspects of military operations, joint training, and logistics necessary to accomplish the missions assigned to the command; should be exercised through the commanders of subordinate organizations; normally exercised through the service or functional component commander; provides full authority to organize and employ commands and forces as the combatant commander considers necessary to accomplish assigned missions (Joint Pub 1-02)
combatant commander	a commander in chief of one of the unified or specified combatant commands established by the President (Joint Pub 1-02)
combat power	the total means of destructive and/or disruptive force which a military unit/formation can apply against the opponent at a given time (Joint Pub 1-02)
combat zone	that area required by combat forces for the conduct of operation; territory forward of the Army rear area boundary (Joint Pub 1-02)
combined action	between two or more forces or agencies of two or more allies (when all allies or services are not involved, the participating nations and services shall be identified (e.g., Combined navies) (Joint Pub 1-02)
combined doctrine	fundamental principles that guide the employment of forces of two or more nations in coordinated action toward a common objective (Joint Pub 1-02)

combined force a military force composed of elements of two or more allied nations (HQ Training and Doctrine Command)

combined operation an operation conducted by forces of two or more allied nations acting together for the accomplishment of a single mission (Joint Pub 1-02)

combined warfare warfare conducted by forces of two or more allied nations in coordinated action toward common objectives (Joint Pub 1-02)

command the authority that a commander in the military service lawfully exercises over subordinates by virtue of rank or assignment; includes the authority and responsibility for effectively using available resources and for planning the employment of, organizing, directing, coordinating, and controlling military forces for the accomplishment of assigned missions (Joint Pub 1-02)

compatibility two or more items or components of equipment or material existing or functioning in the same system or environment without mutual interference

CONPLAN operations plan in concept format

control authority which may be less than full command exercised by a commander over part of the activities of subordinate or other organizations (Joint Pub 1-02)

coordinating authority a commander or individual assigned responsibility for coordinating specific functions or activities involving forces of two or more services or two or more forces of the same service; the commander or individual has the authority to require consultation between the agencies involved, but does not have the authority to compel agreement; in the event that essential agreement cannot be obtained, the matter shall be referred to the appointing authority (Joint Pub 1-02)

COR concept of requirements

CPX command post exercise

CSS combat service support

DCA defensive counterair

delegation of authority action by which a commander assigns part of his authority commensurate with the assigned task to a subordinate commander; while ultimate responsibility cannot be relinquished, delegation of authority carries with it the imposition of a measure of responsibility; the extent of the authority delegated must be clearly stated (Joint Pub 1-02)

doctrine fundamental principles by which the military forces or elements thereof guide their actions in support of national objectives; authoritative but requires judgment in application (Joint Pub 1-02)

DOD Department of Defense

DOS Department of State

DPG defense planning guidance

DSAA Defense Security Assistance Agency

EAC echelons above corps

EPW enemy prisoner of war

EW electronic warfare—any military action involving the use of electromagnetic and directed energy to control the electromagnetic spectrum or to attack the enemy; the three major subdivisions within electronic warfare are electronic attack, electronic protection, and electronic warfare support; in Army usage, use of electromagnetic energy to determine, exploit, reduce, or prevent hostile use of the electromagnetic spectrum and to ensure friendly use thereof

FAA *Foreign Assistance Act*

FAO foreign area officer

FM frequency modulation; field manual

FRAGO fragmentary orders

FSCL fire support coordination line

FSU forward support unit

functional component commander a military commander responsible for the employment and sustainment of assigned forces normally, but not necessarily, composed of forces of two or more services which may be established in peacetime or war to perform particular operational missions that may extend over a period of time (Joint Pub 1-02)

functions the appropriate or assigned duties, responsibilities, missions, or tasks of an individual, office, or organization; as defined in *National Security Act of 1947*, as amended, the term *function* includes functions, powers, and duties (Joint Pub 1-02)

GCC ground component command (Korea only)

GE Germany

GIE global information environment

GO government organization

HET heavy equipment transporter

HN host nation

HNS host nation support—civil and/or military assistance rendered by a nation to foreign forces within its territory during peacetime, crisis or emergencies, or war based upon agreements mutually concluded between nations (Joint Pub 1-02)

HQ headquarters

HUMINT human intelligence

IFF identification, friend or foe

INFOSYS information systems

intelligence the product resulting from the collection, processing, integration, analysis, evaluation, and interpretation of available information concerning foreign countries or areas (Joint Pub 1-02)

interchangeability a condition which exists when two or more items possess such functional and physical characteristics as to be equivalent in performance and durability, and are capable of being exchanged one for the other without alteration of the items themselves or of adjoining items, except for adjustment, and without selection for fit and performance (Joint Pub 1-02)

interoperability the ability of systems, units, or forces to provide services to and accept services from other systems, units, or forces and to use the services so exchanged to enable them to operate effectively together (Joint Pub 1-02)

IO international organization

IPB intelligence-preparation-of-the-battlespace

ISB intermediate staging base

ISE intelligence support element

IT Italy

JAG judge advocate general

JFACC joint force air component commander—derives authority from the JFC who has the authority to exercise operational control, assign missions, direct coordination among subordinate commanders, redirect, and organize forces to ensure unity of effort in the accomplishment of the overall mission; normally designated by the JFC; responsibilities assigned by the joint force commander (normally these include, but are not limited to, planning, coordination, allocation, and tasking based on the JFC's apportionment decision); using the JFC's guidance and authority, and in coordination with other service component commanders and other assigned or supporting commanders, recommends to the JFC apportionment of air sorties to various missions or geographic areas (Joint Pub 1-02)

JFC joint force commander

JFLCC joint force land component commander—the commander within a unified command, subordinate unified command, or joint task force responsible to the establishing commander for making recommendations on the proper employment of land forces, planning and coordinating land operations, or accomplishing such operational missions as may be assigned; JFLCC is given the authority necessary to accomplish missions and tasks assigned by the establishing commander; JFLCC is normally the commander with the preponderance of land forces and the requisite command and control capabilities

JIC joint implementation committee

JIP joint implementation plan

JLP joint logistics plan

JLSP joint logistics support plan

JMC joint movement center

joint connotes activities, operations, organizations, etc., in which elements of more than one service of the same nation participate (when all services are not involved, the participating services shall be identified, e.g., Joint Army-Navy) (Joint Pub 1-02)

joint doctrine fundamental principles that guide the employment of forces of two or more services in coordinated action toward a common objective; promulgated by the Joint Chiefs of Staff, in coordination with the combatant commands, services, and Joint Staff (Joint Pub 1-02)

joint force a general term applied to a force composed of significant elements of the Army, Navy, or Marine Corps and the Air Force or two or more of these services, operating under a single commander authorized to exercise combatant command or operational control over joint forces (Joint Pub 1-02)

JPC/G joint planning commission or group

JSCP joint strategic capabilities plan

JSP joint support plan

JTCB joint targeting coordination board

JTF joint task force

LCC land component commander

lead nation-liaison that contact or intercommunication maintained between elements of military forces to ensure mutual understanding and unity of purpose and action (Joint Pub 1-02)

LNO liaison officer

LOAC laws of armed conflict

LOC line of communication

LOGCAP Logistics Civil Augmentation Program

logistics the science of planning and carrying out the movement and maintenance of forces. In its most comprehensive sense, those aspects of military operations which deal with design and development, acquisition, storage, movement, distribution, maintenance, evacuation, and disposition of material; movement, evacuation, and hospitalization of personnel; acquisition or construction, maintenance, operation, and disposition of facilities; acquisition or furnishing of services (Joint Pub 1-02)

LRC lesser regional contingency

LSE logistics support element

MACOM major Army command

maneuver a movement to place ships or aircraft in a position of advantage over the enemy; a tactical exercise carried out at sea, in the air, on the ground, or on a map in imitation of war; the operation of a ship, aircraft, or vehicle to cause it to perform desired movements; employment of forces on the battlefield through movement in combination with fire, or fire potential, to achieve a position of advantage in respect to the enemy in order to accomplish the mission (Joint Pub 1-02)

MDA multinational deployment agency

MEDEVAC medical evacuation

METT-T mission, enemy, terrain, troops-time available

MFA ministry of foreign affairs

MFC multinational force commander—a general term applied to a commander who exercises command authority over a military force composed of elements of two or more nations; the extent of the MFC's command authority is determined by the participating nations (proposed in Joint Pub 3-16*)*

MIE military information environment

MLS/E multinational support command/element

MLT mobile liaison team

MNF multinational force

MOD ministry of defense

MOOTW military operations other than war—the range of military actions required by the National Command Authorities, except those associated with major combat operations, conducted pursuant to a declaration of war or authorized by the *War Powers Limitation Act*, in support of National Security interests and objectives; these military actions can be applied to complement any combination of the other instruments of national power and occur before, during, and after war. Also known as operations other than war (OOTW) (Joint Pub 3-07)

MRC major regional contingency

MSC/E multinational support command/element

MSR main supply route

multinational operations a collective term to describe military actions conducted by forces of two or more nations, typically organized within the structure of a coalition or alliance (Joint Pub 3-0)

national security objectives those fundamental aims, goals, and purposes of a nation stated in the national security strategy toward which a security policy is directed and efforts and resources of the nation are applied (Joint Pub 3-0)

national security policy a broad course of action or statements of guidance adopted by the government and the national-level command authority in pursuit of national security objectives (Joint Pub 3-0)

National Security Strategy the art and science of developing, applying, and coordinating the instruments of national power (diplomatic, economic, military, and informational) to achieve objectives that contribute to national security (Joint Pub 1-02)

NATO North Atlantic Treaty Organization

NBC warfare nuclear, biological, and chemical warfare

NCA National Command Authorities—the President and the Secretary of Defense or their duly deputized alternates or successors (Joint Pub 1-02)

NDP national disclosure policy

NGO nongovernmental organization—transnational, nonprofit organization of private citizens that maintains a consultative status with the Economic and Social Council of the United Nations; may be professional associations, foundations, multinational businesses, or simply groups with a common interest in humanitarian assistance (development and relief) (Joint Pub 3-07, 16 June 1995)

NSE national support element

objective the physical object of the action taken, e.g., a definite tactical feature, the seizure and/or holding of which is essential to the commander's plan (Joint Pub 1-02)

OCA offensive counterair

OPCOM operational command (NATO)

OPCON operational control—transferable command authority that may be exercised by commanders at any echelon at or below the level of combatant command; inherent in combatant command (command authority) and is the authority to perform those functions of command over subordinate forces involving organizing and employing commands and forces, assigning tasks, designating objectives, and giving authoritative direction necessary to accomplish the mission; includes authoritative direction over all aspects of military operations and joint training necessary to accomplish missions assigned to the command; should be exercised through the commanders of subordinate organizations; normally exercised through the service component commanders; normally provides full authority to organize commands and forces and to employ those forces as the commander in operational control considers necessary to accomplish assigned missions; does not, in and of itself, include authoritative direction for logistics or matters of administration, discipline, internal organization, or unit training (Joint Pub 1-02)

operating forces those forces whose primary missions are to participate in combat and the integral supporting elements thereof (Joint Pub 1-02)

operation a military action or the carrying out of a strategic, tactical, service, training, or administrative military mission; the process of carrying on combat, including movement, supply, attack, defense, and maneuvers needed to gain the objectives of any battle or campaign (Joint Pub 1-02)

operational level of war the level of war at which campaigns and major operations are planned, conducted, and sustained to accomplish strategic objectives within theaters or areas of operations; activities at this level link tactics and strategy by establishing operational objectives needed to accomplish the strategic objectives, sequencing events to achieve the operational objectives, initiating actions, and applying resources to bring about and sustain these events; these activities imply a broader dimension of time or space than do tactics; they ensure the logistics and administrative support of tactical forces and provide the means by which tactical successes are exploited to achieve strategic objectives (Joint Pub 1-02)

OPLAN operation plan

OPORD operations order

OPSEC operations security

PA public affairs

POD port of debarkation

POE port of embarkation

POL petroleum, oils, and lubricants

POLAD political advisor

PSS personnel service support

PSYOP psychological operations—planned operations to convey selected information and indicators to foreign audiences to influence their emotions, motives, objective reasoning, and ultimately the behavior of foreign governments, organizations, groups, and individuals; the purpose is to induce or reinforce foreign attitudes and behaviors favorable to the originator's objectives (Joint Pub 1-02)

PVO private voluntary organization—private, nonprofit humanitarian assistance organization involved in development and relief activities; normally US-based; PVO is often used synonymously with the term *nongovernmental organization* (Joint Pub 3-08, First Working Draft, Dec 94)

QSTAG quadripartite standardization agreement (ABCA)

RC reserve component

RII relevant information and intelligence

ROE rules of engagement

RSI rationalization, standardization, and interoperability

RSO&I reception, staging, onward movement, and integration

RSU rear support unit

SA security assistance

SACEUR Supreme Allied Command Europe

service component command command consisting of the service component commander and all those individuals, units, detachments, organizations, and installations under the command that are assigned to the unified command (Joint Pub 1-02)

SHORAD short-range air defense

SO special operations—operations conducted by specially organized, trained, and equipped military and paramilitary forces to achieve military, political, economic, or psychological objectives by unconventional military means in hostile, denied, or politically sensitive areas; conducted during peacetime competition, conflict, and war, independently or in coordination with operations of conventional, nonspecial operations forces; political-military considerations frequently shape special operations, requiring clandestine, covert, or low visibility techniques and oversight at the national level; differ from conventional operations in degree of physical and political risk, operational techniques, mode of employment, independence from friendly support, and dependence on detailed operational intelligence and indigenous assets (Joint Pub 1-02)

SOF special operations forces

SOFA status of forces agreement

SOP standing operating procedures

SOR statement of requirement

STANAG standardization agreement (NATO)

standardization within NATO, the process of developing concepts, doctrines, procedures, and designs to achieve and maintain the most effective levels of compatibility, interoperability, and commonality in fields of operations, administration, and materiel

strategic level of war the level of war at which a nation, often as a member of a group of nations, determines national or multinational (alliance or coalition) security objectives and guidance and develops and uses national resources to accomplish those objectives; activities at this level establish national and multinational military objectives, sequence initiatives, define limits, and assess risks for the use of military and other instruments of national power; develops global plans or theater war plans to achieve these objectives; provides military forces and other capabilities in accordance with strategic plan (Joint Pub 1-02)

support the action of a force which aids, protects, complements, or sustains another force in accordance with a directive requiring such action; a unit which helps another unit in battle (aviation, artillery, or naval gunfire may be used as a support for infantry); a part of any unit held back at the beginning of an attack as a reserve; an element of a command which assists, protects, or supplies other forces in combat (Joint Pub 1-02)

TA technical agreement or arrangement

TACCIMS Theater Automated Command and Control Information Management System

TACOM tactical command

TACON tactical control—the detailed, and usually, local direction and control of movements or maneuvers necessary to accomplish missions or tasks assigned (Joint Pub 1-02)

tactical level of war the level of war at which battles and engagements are planned and executed to accomplish military objectives assigned to tactical units or task forces; activities at this level focus on the ordered arrangement and maneuver of combat elements in relation to each other and to the enemy to achieve combat objectives (Joint Pub 1-02)

TBM tactical ballistic missile

TCC transportation control center

TMD theater missile defense

TOA transfer of authority

TPFDD time-phased force and deployment data

TTP tactics, techniques, and procedures

TU Turkey

UCP *Unified Command Plan*

UJTL universal joint task list

UK United Kingdom

UN United Nations

unity of command/ unity of effort directing and coordinating the action of all forces toward a common objective; may be achieved by cooperation (the usual method at national and alliance level); at theater level and below, best achieved by vesting a single commander with the requisite authority to direct and to coordinate all forces employed in pursuit of a common goal; at the strategic level, this common goal equates to the political purpose of the United States and the broad strategic objectives which flow from this purpose; at the national level, this common goal determines the military forces necessary for its achievement; the coordination of these forces requires unity of effort; the President as the Commander in Chief of the Armed Forces is assisted in this role by the Secretary of Defense, the Chairman of the Joint Chiefs of Staff, and combatant commands in or in support of a theater; employment of forces in a theater in a manner that develops their full combat power requires unity of effort (Joint Pub 3-0)

UNPROFOR United Nations Protective Force

US United States

USACE US Allied Command Europe

USACOM US Atlantic Command

USAMC US Army Materiel Command

USAREUR US Army European Command

US country team the senior, in-country US coordinating and supervising body, headed by the chief of the US diplomatic mission (usually an ambassador) and composed of the senior member of each represented US department or agency as desired by the chief of the US diplomatic mission (Joint Pub 1-02)

WHNS wartime host nation support

References

SOURCES USED

These are the sources quoted or paraphrased in this publication.

JOINT PUBLICATIONS

Joint Pub 1-02. *Department of Defense Dictionary of Military and Associated Terms.* 23 March 1994.

Joint Pub 3-16. *Joint Doctrine for Multinational Operations* (draft). 30 April 1996.

ARMY PUBLICATIONS

FM 100-5. *Operations.* 14 June 1993.

FM 100-6. *Information Operations.* 27 August 1996.

FM 100-7. *Decisive Force: The Army in Theater Operations.* 31 May 1995.

FM 100-16. *Army Operational Support.* 31 May 1995.

FM 100-23. *Peace Operations.* 30 December 1994.

FM 101-5-1. *Operational Terms and Graphics.* October 1985.

ALLIED PUBLICATIONS

STANAG 2437. *Allied Joint Operations Doctrine* (AJP-1). 15 December 1994.

STANAG 2928. *Land Forces Ammunition Interchangeability Catalogue in Wartime* (AOP-6). 9 June 1995.

STANAG 3680. *NATO Glossary of Terms and Definitions* (AAP-6). 3 October 1986.

DOCUMENTS NEEDED

These documents must be available to the intended users of this publication.

American-British-Canadian-Australian Handbook. December 1995.

JA422 Operational Law Handbook, Judge Advocate General School. 1994

Joint Pub 3-16. *Joint Doctrine for Multinational Operations* (draft). 30 April 1996.

Joint Task Force Commander's Handbook for Peace Operations. 28 February 1995.

Presidential Decision Directive 25 (PPD 25). The Clinton Administration's Policy on Reforming Multilateral Peace Operations. May 1994.

Psychological operations studies that cover many countries around the world; may be obtained by writing to Commander, 4th Psychological Operations Group (Airborne), ATTN: AORC-POG-SB, Fort Bragg, NC 28307-5240.

STANAG 2101/QSTAG 533. *Establishing Liaison.* 28 June 1994.

READINGS RECOMMENDED

These readings contain relevant supplemental information.

JOINT PUBLICATIONS

Joint Pub 0-2. *Unified Action Armed Forces.* 24 February 1995.

Joint Pub 3-0. *Doctrine for Joint Operations.* 1 February 1995.

Joint Pub 3-16. *Joint Doctrine for Multinational Operations* (draft). 30 April 1996.

Joint Pub 4-0. *Doctrine for Logistic Support of Joint Operations.* 27 January 1995.

Joint Pub 4-01.3. *Joint Tactics, Techniques and Procedures for Movement Control.* 21 June 1994.

ARMY PUBLICATIONS

AR 27-50. *Status of Forces Policies, Procedures, and Information.* 15 December 1989.

AR 34-1. *Rationalization, Standardization, and Interoperability Policy.* 15 February 1989.

AR 700-4. *Logistic Assistance Program.* 30 June 1995.

DA Pam 310-35. *Index of International Standardization Agreements.* 15 December 1978.

FM 55-10. *Movement Control in a Theater of Operations.* 8 December 1992.

FM 55-65. *Strategic Deployment.* 3 October 1995.

ALLIED PUBLICATIONS

AAP-4. *NATO Standardization Agreements and Allied Publications.* December 1991.

STANAG 1149. *Doctrine for Amphibious Operations* (ATP-8). 21 December 1995.

STANAG 2014/QSTAG 506. *Operations Orders, Warning Orders, and Administrative/Logistics Orders.* 27 April 1994.

STANAG 2020/QSTAG 510. *Operational Situation Reports.* 14 July 1995.

STANAG 2103. *Reporting Nuclear Detonations Biological and Chemical Attacks, and Predicting and Warning of Associated Hazards and Hazard Areas* (ATP-45). 7 September 1995.

STANAG 2105. *NATO Table of Medical Equivalents* (AMedP-1). 21 September 1989.

STANAG 2394. *Land Force Combat Engineer Doctrine* (ATP-52). 12 January 1993.

STANAG 2406. *Land Forces Logistic Doctrine* (ALP-9[B]). 29 May 1995.

STANAG 2868. *Land Force Tactical Doctrine* (ATP-35[A]). 27 April 1992.

STANAG 2904. *Airmobile Operations* (ATP-41). 12 September 1985.

STANAG 2934. *Artillery Procedures* (AArtyP-1). May 1994.

STANAG 2936. *Intelligence Doctrine* (AIntP-1[A]). 16 May 1995.

STANAG 2999.*Use of Helicopters in Land Operations* (ATP-49[A]). 20 January 1994.

STANAG 3700. *NATO Tactical Air Doctrine* (ATP-33[B]). 27 July 1995.

STANAG 3805. *Doctrine for Airspace Control in Times of Crisis and War* (ATP-40[A]). 9 March 1995.

STANAG 5621. *Standards for the Interoperability of NATO Land Combat and Combined Operations Systems.* 18 December 1989.

STANAG 6010. *Electronic Warfare in the Land Battle* (ATP-51). 12 February 1992.

Index

Made in United States
Troutdale, OR
11/03/2026

56844646R00066